Electrical installation and workshop technology

Electrical installation and workshop technology

Volume one
(Fifth Edition)

F. G. Thompson, I.Eng.FIElecIE, LCG

Senior Lecturer, Department of Technology
Lews Castle College, Stornoway

Illustrative material by

J. H. Smith, AMITE

Late, Dept of Electrical & Electronic Engineering
Inverness College of Further & Higher Education

Longman Scientific & Technical,
Longman Group UK Limited,
Longman House, Burnt Mill, Harlow,
Essex CM20 2JE, England
and Associated Companies throughout the world.

First published 1968
Second edition Metric 1973
Third edition 1978
Fourth edition 1984
Second impression published by Longman Scientific & Technical 1987
Fifth edition 1992

British Library Cataloguing in Publication Data
A catalogue record for this book is available from the British Library

Set in Compugraphic Times 10/12pt

Printed in Malaysia
by Percetakan Mun Sun Sdn. Bhd.,
Shah Alam, Selangor Darul Ehsan

Contents

List of illustrations

Preface

This textbook is intended mainly for students who are studying the subject of electrical installation work. The engineering content of the book is aimed at those whose 'on-the-job' training lacks a significant workshop element.

In this country the electrical contracting industry employs about 60,000 people, virtually all of whom are fully trained and skilled in the types of work that are undertaken by the industry. Many of those now in managerial positions have undergone skills training and technical education which makes the contracting industry one which is highly skilled and able to cope with the increasingly complex demands required by clients, whether they be domestic, commercial or industrial.

Each year some 3,500 apprentices and trainees are recruited by the industry to enter the formal training facilities offered by technical colleges and other training establishments. Success leads to the apprentice being graded as Electrician, Approved Electrician and Technician Electrician; all these grades lead into further promotion opportunities into the managerial structure of the 16,000 reputable companies in the contracting industry, ranging from small family businesses to large sophisticated national companies which service contracts all over the world. The industry is attractive in that it provides full employment for its workforce. Its record for being 'strike free' ensures that the career electrician can look forward to a stable working life full of opportunities for career advancement.

This book caters for those who are following the City & Guilds Course 236, Parts 1 and 2 and, in Scotland, the SCOTVEC Course in Electrical Installation Work, Parts 1 and 2. Both these courses lead to the award of the Electrician's Certificate, the first rung on the career ladder.

The book will also be found useful to those students following the C&G Course 232 in Electrical and Electronic Craft Studies and the C&G Course 823, Overseas Electrical Installation Practice.

Since the first edition of this textbook appeared in 1968 it has been well received by electrical students, teachers and college lecturers, indicating, perhaps, that it fills a need for a practical approach to electrical installation engineering. This new edition thus carries on with the original concept that students entering the career of electrician require a good guide to all the practical aspects of installation work coupled with simple explanations of the more technical aspects of installation technology. Many of these aspects relate to the requirements of the current Regulations for Electrical Installations published by the Institution of Electrical Engineers. They reflect the growing need for safety in electrical work, not just for practising electricians but for those non-electrical persons who come into daily contact, at home or at work, with electrical installations.

The author wishes to thank those who, over the years, have contributed to the book's contents, by suggestions, comments and advice, all of which are now incorporated. Thanks are also due to manufacturers who have given permission for some of the illustrations used in the book.

Part A

Electrical installation technology

1 The career of the electrician

The spectacular increase in the use of electrical energy for virtually all domestic and industrial purposes over the last half century or so is proof enough that the electrical industry plays a most prominent part in the economy of the country. The range of careers which the industry offers is extremely wide. But whatever the career, a fundamental knowledge of electrical engineering, its science and its technology, is necessary for any progress in the career.

The electrician of today plays perhaps one of the most important roles within the electrical industry, and not only in the matter of providing a labour force of skills and abilities on many levels. He is most definitely a key man with a fair degree of responsibility for work which can be carried out satisfactorily only with a background of sound technical knowledge.

Whatever the particular field of employment — supply, manufacturing or contracting — possession of technical knowledge forms the basis for the performance of the many varied tasks which today's electrician is called on by industry and the householder to do — and do well.

Today's electrician works in the supply industry to provide services associated with the generation, transmission and distribution of electrical energy. The 'link-man' between the supply industry and the user of electricity is the electrician employed in the vast field of employment known as the contracting industry. The function of electric wiring is basically to carry electrical energy to a point of use where it is converted into some other form of energy: into light, heat or mechanical power. Wiring, and so the contracting electrician, thus occupies the unique position of being an essential link between the supply authorities on the one hand and the makers and users of all kinds of appliances and apparatus on the other. Once, in years gone by, the idea existed that 'wiring' was hardly a respectable occupation for any but those without technical qualifications. The image has changed, however, and the electrician of today is required to have a minimum certificate to indicate his ability to do his job to a certain standard. Indeed, in some countries overseas, it is impossible for anyone to set up a career as an electrician unless he has passed a theoretical and practical examination: this is a legal requirement.

Electricity in this country, indeed in any country, is much more than a national asset. It has a deep social importance. It has been the main influence for good in the field of improvement in our standards of living. Electricity has proved to be the most flexible form of power in existence: it can be generated easily and transmitted to whatever or whomever requires it, and in whatever quantity it may be required. The historian Trevelyan has said that the social scene grows out of economic conditions. It is true to say that with the aid of electricity, a livelihood and a way of life without drudgery has been won, with an attendant enrichment of the social scene. That electricity is now commonplace in our lives is proved by the fact that the ordinary user regards it as essential to his daily life and living, and his work, and accepts it completely without question.

Offering, as it does to the user, a clean and effortless way of life, electricity also offers to the electrician a means whereby an interesting and satisfying living can be made. The opportunities available are legion, but particularly to those who qualify for promotion by study. Emphases nowadays placed on the possession of some minimum qualifications are so insistent that progress in any electrical career is virtually impossible without it. But theory alone is not enough. Practical experience is also a vital necessity; for not only must a job be done, but it must be understood thoroughly how and why it is done.

So far as the electrician of today is concerned, the bulk of the practical training is through the medium of an apprenticeship lasting four years. This period is very important. For it offers not only a means of earning some money while training for a career, but is very much like a foundation-stone for future progress in that career. Faulty instruction, for instance, may mar the chances which come up in later life, and may indeed leave a recognisable mark — that of the incompetent tradesman. Attention to all approved instruction received during the apprenticeship period will therefore ensure that the career progresses surely and steadily to attain whatever positions are offered within the industry.

Technical knowledge and experience are thus essentials which form a background to responsibility. In the contracting industry, the delegation of responsibility is a necessary feature of any class of work undertaken. On a building site, for instance, one can see the various grades of responsibility handed over to qualified electricians: chargehand, foreman, site engineer and so on. Responsibility is not only for the quality of performance of the men being supervised, but also for the quality of the materials used in the contract and how they are applied.

In the manufacturing industry, responsibility is given to those who have proved themselves able to ensure, by applying their knowledge and experience, that electrical energy is used in the most efficient way, so increasing productivity in the application and maintenance aspects.

The basic requirement for any worthwhile career in the electrical industry is the possession of a relevant qualification, obtained through a technical college or other training agency. This qualification is offered by the City & Guilds and, in Scotland, the Scottish Vocational Education Council. Both these bodies oversee the technical education requirements which are the foundation of the electrician's knowledge of his work. Then there is the practical training, obtained through 'on-site' activity and skills training in the colleges. Success and proficiency shown by the student are rewarded by he or she being recognised as an Electrician, Approved Electrician and, later for those with ambition, Technician Electrician.

All new entrants are required to be physically fit and healthy and must not be colour-blind. The educational training is quite demanding, students having to cope with complex technical subjects in technology and electrical principles. Usually the minimum entry requirements are good grades in English, maths and science. However the industry does accept students who have no formal school qualifications but who show in selection and aptitude tests a dedication and commitment to becoming an electrician.

Once the student has been accepted as an apprentice, the sponsoring employer enters into a training contract which requires the student to undertake the necessary education and training. The contract also guarantees the student employment for the whole of the training period. Contracts are registered with the Joint Industry Board, the JIB (in Scotland the SJIB), which is a special organisation set up by the employers and the union, the EETPU. The student then becomes a junior apprentice and is required to spend a minimum of twenty-four weeks at a local technical college for technical education and practical installation work. The remainder of the first year is spent with the employer for training on site, thus giving the student the opportunity to enhance skills and to apply theoretical knowledge.

At the end of the first year the student is entered for the C&G Course 236 (Part 1) examination in Electrical Installation Work and takes a practical test known as 'Achievement Measurement Test 1' (AM1). In Scotland the student training programme, also over a period of twenty-four weeks, consists of Modules, each Module dealing with a specific subject (e.g. Inspection and Testing), the study of which can take either twenty or forty hours to complete. As each Module is achieved, through a system of tests called Assessments, the student builds up a number of credits which, over the twenty-four weeks, results in a SCOTVEC First Stage Certificate. The Assessments can be either 'paper' tests or practical skills tests (e.g. Cable Tray Systems).

Success in the first year attracts a pay increase, always a useful incentive to the student to do his or her best to achieve entry to the second year of study. This second year involves a further period of study at a technical college to prepare for, in England and Wales, the City & Guilds 236 Part 2

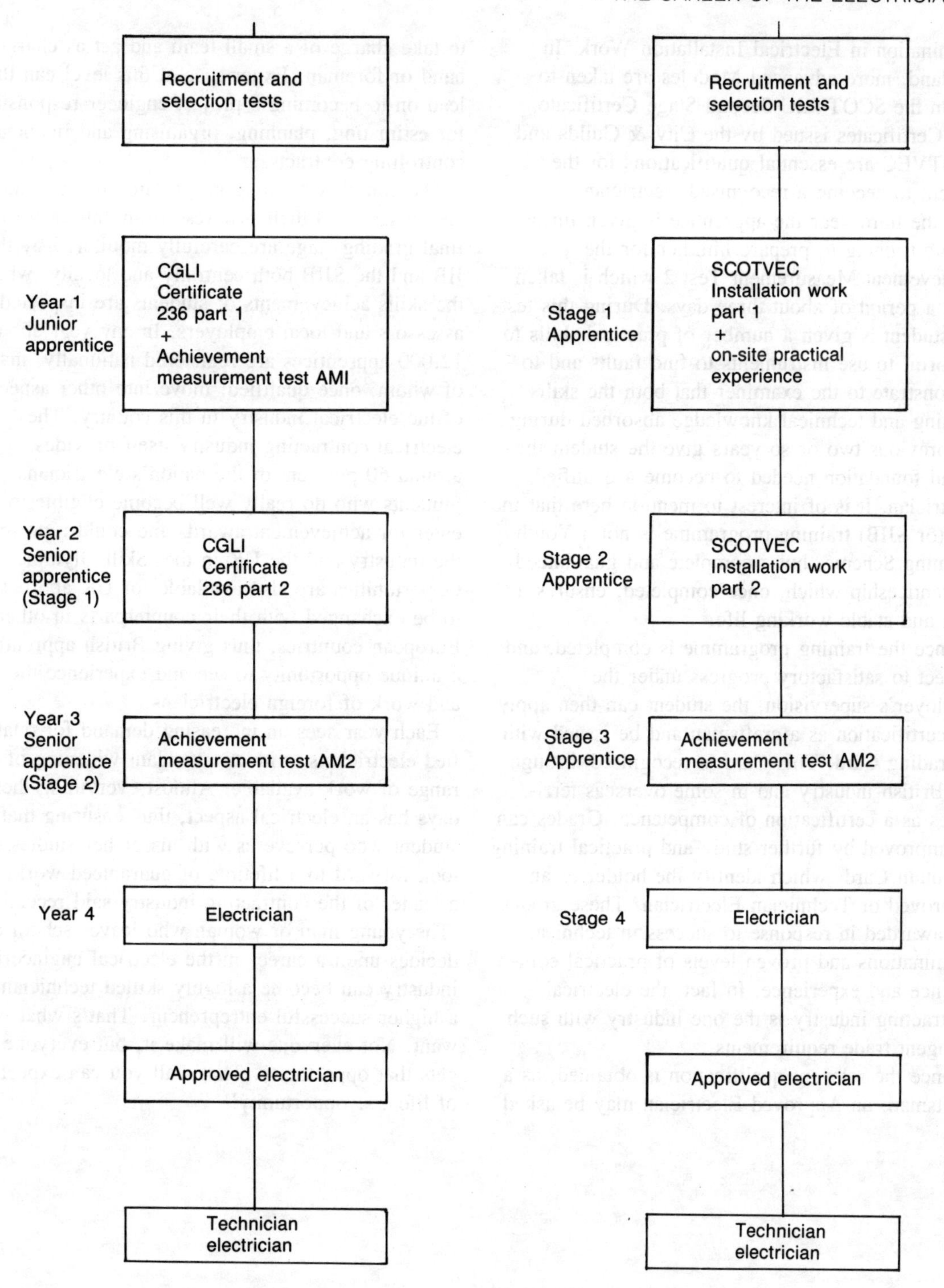

Figure 1.1. The routes through the training system for apprentice electricians. The right-hand side applies to Scotland only.

examination in Electrical Installation Work. In Scotland, more advanced Modules are taken to obtain the SCOTVEC Second Stage Certificate. The Certificates issued by the City & Guilds and SCOTVEC are essential qualifications for the student to become a recognised electrician.

In the third year the apprentice is given further on-site training to prepare him/her for the Achievement Measurement Test 2 which is taken over a period of about three days. During this test the student is given a number of practical skills to perform, to use instruments to find faults and to demonstrate to the examiner that both the skills training and technical knowledge absorbed during the previous two or so years give the student the sound foundation needed to become a qualified electrician. It is of interest to mention here that the JIB (or SJIB) training programme is not a Youth Training Scheme, but a complete and guaranteed apprenticeship which, once completed, ensures a long and stable working life.

Once the training programme is completed, and subject to satisfactory progress under the employer's supervision, the student can then apply for certification as a craftsman and be issued with a Grading Card. This Card is recognised throughout British industry and in some overseas territories as a certification of competence. Grades can be improved by further study and practical training to obtain Cards which identify the holder as an Approved or Technician Electrician. These grades are awarded in response to success in technical examinations and proven levels of practical competence and experience. In fact, the electrical contracting industry is the one industry with such stringent trade requirements.

Once the relevant qualification is obtained, as a craftsman, an Approved Electrician may be asked to take charge of a small team and act as chargehand or foreman. Experience at this level can then lead on to becoming a project engineer responsible for estimating, planning, organising and financially controlling contracts.

The complete training programme for apprentice electricians and their progress from entrant to the final grading stage are carefully monitored by the JIB and the SJIB both centrally and locally, where the skills achievements of students are inspected by assessors and local employers. In any year about 12,000 apprentices are registered nationally, many of whom, once qualified, move into other aspects of the electrical industry in this country. The electrical contracting industry itself provides around 60 per cent of the nation's electricians. Students who do really well become eligible to enter for achievement awards and could represent the industry and the UK in the 'Skill Olympics'. Opportunities are also available for UK apprentices to be exchanged with their counterparts in other European countries, thus giving British apprentices a unique opportunity to see and experience the life and work of foreign electricians.

Each year sees an increasing demand for qualified electricians, with an attendant widening of the range of work available. Almost everything these days has an electrical aspect, thus ensuring that the student who perseveres with his or her studies can look forward to a lifetime of guaranteed work. As a leader of the contracting industry said recently: 'The young man or woman who leaves school and decides upon a career in the electrical engineering industry can become a highly skilled technician or a higher successful entrepreneur. That's what we want. Not everyone will make it, but everyone gets that opportunity. That's all you can expect out of life . . . opportunity!'

2 Historical review of installation work

As one might expect to find in the early beginnings of any industry, the application, and the methods of application, of electricity for lighting, heating and motive power was primitive in the extreme. Large-scale application of electrical energy was slow to develop. The first wide use of it was for lighting in houses, shops and offices. By the 1870s, electric lighting had advanced from being a curiosity to something with a definite practical future. Arc lamps were the first form of lighting, particularly for the illumination of main streets. When the incandescent-filament lamp appeared on the scene electric lighting took on such a prominence that it severely threatened the use of gas for this purpose. But it was not until cheap and reliable metal-filament lamps were produced that electric lighting found a place in every home in the land. Even then, because of the low power of these early filament lamps, shop windows continued for some time to be lighted externally by arc lamps suspended from the fronts of buildings.

The earliest application of electrical energy as an agent for motive power in industry is still electricity's greatest contribution to industrial expansion. The year 1900 has been regarded as a time when industrialists awakened to the potential of the new form of power.

Electricity was first used in mining for pumping. In the iron and steel industry, by 1917, electric furnaces of both the arc and induction type were producing over 100,000 tons of ingots and castings. The first all-welded ship was constructed in 1920; and the other ship-building processes were operated by electric motor power for punching, shearing, drilling machines and woodworking machinery.

The first electric motor drives in light industries were in the form of one motor-unit per line of shafting. Each motor was started once a day and continued to run throughout the whole working day in one direction at a constant speed. All the various machines driven from the shafting were started, stopped, reversed or changed in direction and speed by mechanical means. The development of integral electric drives, with provisions for starting, stopping and speed changes, led to the extensive use of the motor in small kilowatt ranges to drive an associated single machine, e.g. a lathe. One of the pioneers in the use of motors was the firm of Bruce Peebles, Edinburgh. The firm supplied, in the 1890s, a number of weatherproof, totally-enclosed motors for quarries in Dumfriesshire, believed to be among the first of their type in Britain. The first electric winder ever built in Britain was supplied in 1905 to a Lanark oil concern. Railway electrification started as long ago as 1883, but it was not until long after the turn of this century that any major development took place.

Electrical installations in the early days were quite primitive and often dangerous. It is on record that in 1881, the installation in Hatfield House was carried out by an aristocratic amateur. That the installation was dangerous did not perturb visitors to the house who '. . . when the naked wires on the gallery ceiling broke into flame . . . nonchalantly threw up cushions to put out the fire and then went on with their conversation'.

Many names of the early electrical pioneers survive today. Julius Sax began to make electric bells in 1855, and later supplied the telephone with which Queen Victoria spoke between Osborne, in the Isle of Wight, and Southampton in 1878. He founded one of the earliest purely electrical manufacturing firms which exists today and still makes bells and signalling equipment.

The General Electric Company had its origins in the 1880s, as a Company which was able to supply every single item which went to form a complete electrical installation. In addition it was guaranteed that all the components offered for sale were technically suited to each other, were of adequate quality and were offered at an economic price.

Specialising in lighting, Falk Stadelmann & Co. Ltd began by marketing improved designs of oil lamps, then gas fittings, and ultimately electric lighting fittings.

Cable makers W. T. Glover & Co. were pioneers in the wire field. Glover was originally a designer of textile machinery, but by 1868 he was also making braided steel wires for the then fashionable crinolines. From this type of wire it was a natural step to the production of insulated conductors for electrical purposes. At the Crystal Palace Exhibition in 1885 he showed a great range of cables; he was also responsible for the wiring of the exhibition.

The well-known J. & P. firm (Johnson & Phillips) began with making telegraphic equipment, extended to generators and arc lamps, and then to power supply.

The coverings for the insulation of wires in the early days included textiles and gutta-percha. Progress in insulation provisions for cables was made when vulcanised rubber was introduced, and it is still used today. The first application of a lead sheath to rubber-insulated cables was made by Siemens Brothers. The manner in which we name cables was also a product of Siemens, whose early system was to give a cable a certain length related to a standard resistance of 0.1 ohm. Thus a No. 90 cable in their catalogue was a cable of which 90 yards had a resistance of 0.1 ohm. Cable sizes were also generally known by the Standard Wire Gauge.

For many years ordinary VRI cables made up about 95 per cent of all installations. They were used first in wood casing, and then in conduit. Wood casing was a very early invention. It was introduced to separate conductors, this separation being considered a necessary safe-guard against the two wires touching and so causing fire. Choosing a cable at the turn of the century was quite a task. From one catalogue alone, one could choose from fifty-eight sizes of wire, with no less than fourteen different grades of rubber insulation. The grades were described by such terms as light, high, medium or best insulation. Nowadays there are two grades of insulation: up to 600 V and 600 V/1,000 V. And the sizes of cables have been reduced to a more practicable seventeen.

During the 1890s the practice of using paper as an insulating material for cables was well established. One of the earliest makers was the company which later became a member of the present-day BICC Group. The idea of using paper as an insulation material came from America to Britain where it formed part of the first wiring system for domestic premises. This was twin lead-sheathed cables. Bases for switches and other accessories associated with the system were of cast solder, to which the cable sheathing was wiped, and then all joints sealed with a compound. The compound was necessary because the paper insulation when dry tends to absorb moisture.

In 1911, the famous 'Henley Wiring System' came on the market. It comprised flat-twin cables with a lead-alloy sheath. Special junction boxes, if properly fixed, automatically effected good electrical continuity. The insulation was rubber. It became very popular. Indeed, it proved so easy to install that a lot of unqualified people appeared on the contracting scene as 'electricians'. When it received the approval of the IEE Rules, it became an established wiring system and is still in use today.

At the time the lead-sheathed system made its first appearance, another rival wiring system also came onto the scene. This was the CTS system (cab-tyre sheathed). It arose out of the idea that if a rubber product could be used to stand up to the wear and tear of motor-car tyres on roads, then the material would well be applied to cover cables. The CTS name eventually gave way to TRS (tough-rubber sheath), when the rubber-sheathed cable system came into general use.

The main competitor to rubber as an insulating material appeared in the late 1930s. This material was PVC (polyvinylchloride), a synthetic material which came from Germany. The material, though inferior to rubber so far as elastic properties were concerned, could withstand the effects of both oil and sunlight. During the Second World War PVC, used both as wire insulation and the protective sheath, became well established.

As experience increased with the use of TRS

cables, it was made the basis of modified wiring systems. The first of these was the Callender farm-wiring system introduced in 1937. This was tough-rubber sheathed cables with a semi-embedded braiding treated with a green-coloured compound. This system combined the properties of ordinary TRS and HSOS (house-service overhead system) cables.

So far as conductor material was concerned, copper was the most widely used. But aluminium was also applied as a conductor material. Aluminium, which has excellent electrical properties, has been produced on a large commercial scale since about 1890. Overhead lines of aluminium were first installed in 1898. Rubber-insulated aluminium cables of 3/0.036 inch and 3/0.045 inch were were made to the order of the British Aluminium Company and used in the early years of this century for the wiring of the staff quarters at Kinlochleven in Argyllshire. Despite the fact that lead and lead-alloy proved to be of great value in the sheathing of cables, aluminium was looked to for a sheath of, in particular, light weight. Many experiments were carried out before a reliable system of aluminium-sheathed cable could be put on the market.

Perhaps one of the most interesting systems of wiring to come into existence was the MICS (mineral-insulated copper-sheathed cable) which used compressed magnesium oxide as the insulation, and had a copper sheath and copper conductors. The cable was first developed in 1897 and was first produced in France. It has been made in Britain since 1937, first by Pyrotenax Ltd, and later by other firms. Mineral insulation has also been used with conductors and sheathing of aluminium.

One of the first suggestions for steel used for conduit was made in 1883. It was then called 'small iron tubes'. However, the first conduits were of bitumised paper. Steel for conduits did not appear on the wiring scene until about 1895. The revolution in conduit wiring dates from 1897, and is associated with the name 'Simplex' which is common enough today. It is said that the inventor, L. M. Waterhouse, got the idea of close-joint conduit by spending a sleepless night in a hotel bedroom staring at the bottom rail of his iron bedstead. In 1898 he began the production of light gauge close-joint conduits. A year later the screwed-conduit system was introduced.

Non-ferrous conduits were also a feature of the wiring scene. Heavy-gauge copper tubes were used for the wiring of the Rylands Library in Manchester in 1886. Aluminium conduit, though suggested during the 1920s, did not appear on the market until steel became a valuable material for munitions during the Second World War.

Insulated conduits also were used for many applications in installation work, and are still used to meet some particular installation conditions. The 'Gilflex' system, for instance, makes use of a PVC tube which can be bent cold, compared with earlier material which required the use of heat for bending.

Accessories for use with wiring systems were the subject of many experiments; many interesting designs came onto the market for the electrician to use in his work. When lighting became popular, there arose a need for the individidual control of each lamp from its own control point. The 'branch switch' was used for this purpose. The term 'switch' came over to this country from America, from railway terms which indicated a railway 'point', where a train could be 'switched' from one set of tracks to another. The 'switch', so far as the electric circuit was concerned, thus came to mean a device which could switch an electric current from one circuit to another.

It was Thomas Edison who, in addition to pioneering the incandescent lamp, gave much thought to the provision of branch switches in circuit wiring. The term 'branch' meant a tee-off from a main cable to feed small current-using items. The earliest switches were of the 'turn' type, in which the contacts were wiped together in a rotary motion to make the circuit. The first switches were really crude efforts: made of wood and with no positive ON or OFF position. Indeed, it was usual practice to make an inefficient contact to produce an arc to 'dim' the lights! Needless to say, this misuse of the early switches, in conjunction with their wooden construction, led to many fires. But new materials were brought forward for switch construction such as slate, marble, and, later, porcelain. Movements were also made more positive with definite ON and OFF positions.

The 'turn' switch eventually gave way to the 'tumbler' switch in popularity. It came into regular use about 1890. Where the name 'tumbler' originated is not clear; there are many sources, including the similarity of the switch action to the antics of Tumbler Pigeons. Many accessory names which are household words to the electricians of today appeared at the turn of the century: Verity's, McGeoch, Tucker and Crabtree. Further developments to produce the semi-recessed, the flush, the ac only, and the 'silent' switch proceeded apace. The switches of today are indeed of long and worthy pedigrees.

It was one thing to produce a lamp operated from electricity. It was quite another thing to devise a way in which the lamp could be held securely while current was flowing in its circuit. The first lamps were fitted with wire tails for joining to terminal screws. It was Thomas Edison who introduced, in 1880, the screw-cap which still bears his name. It is said he got the idea from the stoppers fitted to kerosene cans of the time. Like many another really good idea, it superseded all its competitive lampholders and its use extended through America and Europe. In Britain, however, it was not popular. The bayonet-cap type of lampholder was introduced by the Edison & Swan Co. about 1886. The early type was soon improved to the lampholders we know today.

Ceiling roses, too, have an interesting history; some of the first types incorporated fuses. The first rose for direct attachment to conduit came out in the early 1900s, introduced by Dorman & Smith Ltd.

The first patent for a plug-and-socket was brought out by Lord Kelvin, a pioneer of electric wiring systems and wiring accessories. The accessory was used mainly for lamp loads at first, and so carried very small currents. However, domestic appliances were beginning to appear on the market, which meant that sockets had to carry heavier currents. Two popular items were irons and curling-tong heaters. Shuttered sockets were designed by Crompton in 1893. The modern shuttered type of socket appeared as a prototype in 1905, introduced by 'Diamond H'. Many sockets were individually fused, a practice which was later extended to the provision of a fuse in the plug. These fuses were, however, only a small piece of wire between two terminals and caused such a lot of trouble that in 1911 the Institution of Electrical Engineers banned their use. One firm which came into existence with the socket-and-plug was M.K. Electric Ltd. The initials were for 'Multi-Kontakt' and associated with a type of socket-outlet which eventually became the standard design for this accessory. It was Scholes, under the name of 'Wylex', who introduced a revolutionary design of plug-and-socket: a hollow circular earth pin and rectangular current-carrying pins. This was really the first attempt to 'polarise', or to differentiate between live, earth and neutral pins.

One of the earliest accessories to have a cartridge fuse incorporated in it was the plug produced by Dorman & Smith Ltd. The fuse actually formed one of the pins, and could be screwed in or out when replacement was necessary. It is a rather long cry from those pioneering days to the present system of standard socket-outlets and plugs.

Early fuses consisted of lead wires, lead being used because of its low melting point. Generally, devices which contained fuses were called 'cutouts', a term still used today for the item in the sequence of supply-control equipment entering a building. Once the idea caught on of providing protection for a circuit in the form of fuses, brains went to work to design fuses and fusegear. Control gear first appeared encased in wood. But ironclad versions made their due appearance, particularly for industrial use during the nineties. They were usually called 'motor switches', and had their blades and contacts mounted on a slate panel. Among the first companies in the switchgear field were Bill & Co., Sanders & Co. and the MEM Co., whose 'Kantark' fuses are so well known today. In 1928 this Company introduced the 'splitter' which effected a useful economy in many of the smaller installations.

It was not until the 1930s that the distribution of electricity in buildings by means of busbars came into fashion, though the system had been used as far back as about 1880, particularly for street mains. In 1935 the English Electric Co. introduced a busbar trunking system designed to meet the needs of the motor-car industry. It provided the overhead distribution of electricity into which system individual machines could be tapped wherever required; this idea caught on and designs were produced and put onto the market by Marryat & Place, GEC and Ottermill.

Trunking came into fashion mainly because the larger sizes of conduit proved to be expensive and troublesome to instal. One of the first trunking types to be produced was the 'spring conduit' of the Manchester firm of Key Engineering. They showed it for the first time at an electrical exhibition in 1908. It was semi-circular steel troughing with edges formed in such a way that they remained quite secure by a spring action after being pressed into contact. But it was not until about 1930 that the idea took root and is now established as a standard wiring system.

The story of electric wiring, its systems and accessories tells an important aspect in the history of industrial development and in the history of social progress. The inventiveness of the old electrical personalities, Crompton, Swan, Edison, Kelvin and many others, is well worth noting; for it is from their brain-children that the present-day electrical contracting industry has evolved to become one of the most important sections of activity in electrical engineering. For those who are interested in details of the evolution and development of electric wiring systems and accessories, good reading can be found in the book by J. Mellanby: *The History of Electric Wiring* (MacDonald, London).

Any comparison of manufacturers' catalogues of, say, ten years ago, with those of today will quickly reveal how development of both wiring systems and wiring accessories have changed, not only physically, in their design and appearance but in their ability to meet the demands made on them of modern electrical installations, both domestic and industrial. What were once innovations, such as dimmer switches, for instance, are now fairly commonplace where clients require more flexible control of domestic circuits. The new requirements of the Regulations for Electrical Installations will no doubt introduce more changes in wiring systems and accessories so that installations become safer to use with attendant reductions in the risk from electric shock and fire hazards. New developments in lighting, for instance, particularly during the last decade or so, herald changes in the approach to installation work. Innovative changes in space and water heating, using solar energy and heat pumps, will involve the electrician in situations which can offer exciting challenges in installation work, not least in keeping up with the new face of old technology. More and more is the work of the electrician becoming an area of activity where a thorough grasp of the technology involved is essential if one is to offer the client a safe, reliable and technically competent installation.

3 Historical review of wiring regulations

The history of the development of non-legal and statutory rules and regulations for the wiring of buildings is no less interesting than that of wiring systems and accessories. When electrical energy received a utilisation impetus from the invention of the incandescent lamp, many set themselves up as electricians or electrical wiremen.* Others were gas plumbers who indulged in the installation of electrics as a matter of normal course. This was all very well: the contracting industry had to get started in some way, however ragged. But with so many amateurs troubles were bound to multiply. And they did. It was not long before arc lamps, sparking commutators, and badly insulated conductors contributed to fires. It was the insurance companies which gave their attention to the fire risk inherent in the electrical installations of the 1880s. Foremost among these was the Phoenix Assurance Co., whose engineer, Mr Heaphy, was told to investigate the situation and draw up a report on his findings.

The result was the Phoenix Rules of 1882. These Rules were produced just a few months after those of the American Board of Fire Underwriters who are credited with the issue of the first wiring rules in the world.

The Phoenix Rules were, however, the better set and went through many editions before revision was thought necessary. That these Rules contributed to a better standard of wiring, and introduced a high factor of safety in the electrical wiring and equipment of buildings, was indicated by a report in 1892 which showed the high incidence of electrical fires in the USA and the comparative freedom from fires of electrical origin in Britain.

Three months after the issue of the Phoenix Rules for wiring in 1882, the Society of Telegraph Engineers and Electricians (now the Institution of Electrical Engineers) issued the first edition of *Rules and Regulations for the Prevention of Fire Risks arising from Electric Lighting*. These rules were drawn up by a committee of eighteen men which included some of the famous names of the day: Lord Kelvin, Siemens and Crompton. The Rules, however, were subjected to some criticism. Compared with the Phoenix Rules they left much to be desired. But the Society was working on the basis of laying down a set of principles rather than, as Heaphy did, drawing up a guide or 'Code of Practice'. A second edition of the Society's Rules was issued in 1888. The third edition was issued in 1897 and entitled *General Rules recommended for Wiring for the Supply of Electrical Energy*.

The Rules have since been revised at fairly regular intervals as new developments and the results of experience can be written in for the considered attention of all those concerned with the electrical equipment of buildings. Basically the rgulations were intended to act as a guide for electricians and others to provide a degree of safety in the use of electricity by inexperienced persons such as householders. The regulations were, and still are, not legal; that is, they cannot be enforced by the law of the land. Despite this apparent loophole, the regulations are accepted as a guide to the practice of installation work which will ensure, at the very least, a minimum standard of work. The Institution of Electrical Engineers (IEE) was not alone in the insistence of good standards in electrical installation work. In 1905, the Electrical Trades Union, through the London District Committee, in a letter to the Phoenix Assurance Co., said ' . . . they view with alarm the large extent to which bad work is now being carried out by electric light contractors . . . . As the carrying out of bad work is attended by fires and other risks, besides injuring the Trade, they respectfully ask

* The first wiremen were originally carpenters or bell-hangers, who instructed in the 'new system of electric lighting'.

you to . . . uphold a higher standard of work'.

The legislation embodied in the Factory and Workshop Acts of 1901 and 1907 had a considerable influence on wiring practice. In the latter Act it was recognised for the first time that the generation, distribution and use of electricity in industrial premises could be dangerous. To control electricity in factories and other premises a draft set of Regulations was later to be incorporated into Statutory requirements.

While the IEE and the Statutory regulations were making their positions stronger, the British Standards Institution brought out, and is still issuing, Codes of Practice to provide what are regarded as guides to good practice. The position of the Statutory Regulations in this country is that they form the primary requirements which must by law be satisfied. The IEE Regulations and Codes of Practice indicate supplementary requirements. However, it is accepted that if an installation is carried out in accordance with the IEE Wiring Regulations, then it generally fulfils the requirements of the Electricity Supply Regulations. This means that a supply authority can insist upon all electrical work to be carried out to the standard of the IEE Regulations, but cannot insist on a standard which is in excess of the IEE requirements.

The position of the IEE 'Regs', as they are popularly called, is that of being the installation engineer's 'bible'. Because the Regulations cover the whole field of installation work, and if they are complied with, it is certain that the resultant electrical installation will meet the requirements of all interested parties. There are, however, certain types of electrical installations which require special attention to prevent fires and accidents. These include mines, cinemas, theatres, factories and places where there are exceptional risks.

The following list gives the principal regulations which cover electricity supply and electrical installations:

Non-Statutory Regulations:

1. Institute of Electrical Engineers Regulations for Electrical Installations — This covers industrial and domestic electrical installation work in buildings.
2. The Institute of Petroleum Electrical Code, 1963 — This indicates special safety requirements in the petroleum industry, including protection from lightning and static. It is supplementary to the IEE Regulations.
3. Factories Act, 1961. Memorandum by the Senior Electrical Inspector of Factories — Deals with installations in factories.
4. Explanatory Notes on the Electricity Supply Regulations, 1937 — These indicate the requirements governing the supply and use of electricity.
5. Hospital Technical Memoranda No. 7 — Indicates the electrical services, supply and distribution in hospitals.

All electrical contractors are most particularly concerned with the various requirements laid down by Acts of Parliament (or by Orders and Regulations made thereunder) as to the method of installing electric lines and fittings in various premises, and as to their qualities and specifications.

Statutory Regulations:

1. Building (Scotland) Act, 1959 — Provides for minimum standards of construction and materials including electrical installations.
2. Building Standards (Scotland) Regulations, 1981 — Contains minimum requirements for electrical installations.
3. Electricity Supply Regulations, 1937 — Indicates the requirements governing the supply and use of electricity and deals with installations generally, subject to certain exemptions.
4. Electricity (Factories Act) Special Regulations, 1908 and 1944 — Deals with factory installations, installations on construction sites, and installations of non-domestic caravans such as mobile workshops. These Regulations come under the authority of the Health and Safety Commission.
5. Coal and other Mines (Electricity) Regulations, 1956 — Deals with coal-mine installations.
6. Cinematograph (Safety) Regulations, 1952 — Deals with installations in cinemas.

7. Quarries (Electricity) Regulations, 1956 — Deals with installations at quarry operations.
8. Agriculture (Stationary Machinery) Regulations, 1959 — Deals with agricultural and horticultural installations.

Though these Statutory Regulations are concerned with electrical safety in the respective type of installations listed, there are other Statutory Regulations which are also concerned with electrical safety when equipment and appliances are being used. Included in these are the Electricity at Work Regulations which came into force in 1990. They are stringent in their requirements that all electrical equipment used in schools, colleges, factories and other places of work is in a safe condition and must be subjected to regular testing by competent persons.

Because of the rather legal language in which many of the Statutory Regulations are written, a number of them are made the subject of Guides and Explanatory Notes so that the electrical contractor and his employees are better able to understand the requirements.

It should be noted that in addition to the above list, there are quite a number of Statutory Regulations which deal with specific types of installations such as caravans and petrol stations. While it may seem that the electrician is completely surrounded by Regulations, it should be remembered that their purpose is to ensure not only the safety of the public, but work persons also. And it is also worth noting that in the UK the record for the lowest number of electrical accidents is among the best in the world.

It is a requirement of the current edition of the IEE Regulations for Electrical Installations that good workmanship and the use of approved materials contribute to the high level of safety provided in any electrical installation. The British Standards Institution (BSI) is the approved body for the preparation and issue of Standards for testing the quality of materials and their performance once they are installed in buildings. A typical Standard is BS 31 *Steel Conduit and Fittings for Electrical Wiring*. The BSI also issues Codes of Practice which indicate acceptable standards of good practice and take the form of recommendations. These Codes contain the many years of practical experience of electrical contractors. Some of the Codes of interest to the practising electrician include:

BS 1003: Electrical apparatus and associated equipment for use in explosive atmospheres of gas or vapour
BS 7375: Distribution of electricity on construction and building sites
BS 1018: Electric floor-warming systems for use with off-peak and similar supplies of electricity

Almost a century after the first Wiring Regulations were issued a complete revision was made in 1981 with the appearance of the 15th edition under the title *Regulations for Electrical Installations*. This edition differed from previous editions in its highly technical approach to the provision of electrical installations, based on the need for a high degree of quality of both materials and workmanship to ensure safety from fire, shock and burns. The technical content of the 15th edition of the Regulations placed a degree of responsibility on practising electricians to become familiar with the electrical science principles and the technology which the installer must have in order to provide a client with an installation which is well designed and safe to use.

The 16th edition is now published with yet more changes and differences in approach from the 15th edition. The major changes include the smaller number of explanatory notes and fewer appendices. The 16th edition is also accompanied by a number of other publications: *Guidance Notes* and an *On-site Guide*. The *Guidance Notes* give detailed information on such topics as protection against electric shock, protection against overcurrent, initial and periodic testing and special installations and locations. The *On-site Guide* provides guidance on the construction of the smaller installation such as domestic, commercial and small three-phase installations without the need for the considerable amount of calculations which the 15th edition required in the design of an installation. The Guide in fact offers information which will ensure that an installation has a high degree of built-in safety without taking economic cost into consideration. The Guide also contains much 'need-to-know' information, thus making the technical aspects of an electrical installation more accessible to the practising electrician.

In short, the new 16th edition of the Regulations still places responsibility on the electrician to fully understand the technical aspects of the work he carries out which is only to be expected from a skilled and qualified work person.

While the IEE Wiring Regulations have, since 1882, become a widely recognised standard for electrical installations, they have not had any legal status, except when they are quoted for contractual purposes. With the creation of the Single Common Market and the harmonisation of, among many other things, electrical standards among the member countries of the Common Market, the Regulations, from 1992, have been given an enhanced status by being allotted a British Standard number.

4 Electricity supply and distribution systems

Though there are many forms of energy, it is generally accepted that electricity is by far the most convenient and flexible. The bulk of electricity generated in this country is obtained by burning a low-grade coal to produce high-pressure steam. This steam is then fed to machines known as steam-turbines, which are mechanically coupled to electric generators. For reasons of economy in fuel, plant and manpower electricity is generated in extremely large power stations which are sited throughout the country at places as close as possible to where the highest load-demands occur.

The electric generators or alternators are three-phase machines generating alternating current. As the name implies, alternating current refers to a current whose direction of flow is first in one direction and then in an opposite direction, these alternations occurring at a regular period or frequency. One complete alternation is called a 'hertz'. The frequency of British mains is standardised at 50 hertz (Hz).

In certain parts of Britain, where there are large amounts of water-power, the water is harnessed to drive water-turbines. The site of a hydroelectric power station is very much determined by natural conditions, which are usually found to be most convenient quite long distances away from the heavy load centres. For the efficient operation of hydroelectric plant, there must be water available at a usable head (height above sea-level) and in sufficient quantity. And if the flow of water is not regular enough for a continuous supply for the turbines, then accommodation for water storage must be provided by building dams. The water from the dams is led through large-diameter pipes to the power stations.

Scotland and some parts of the mountainous areas of Wales have sufficient water-power resources for hydroelectric schemes. The bulk of hydroelectric generation is, however, in central and north Scotland. The turbines, as in steam power stations, are mechanically coupled to the electric generators.

The most modern method of generation is by means of nuclear power, in which the heat produced by nuclear fission is used to raise steam, which is then fed to steam turbines to drive the electric generators.

Other methods of generation in Britain include the diesel engine, very common for private standby plants, and in areas such as the small Scottish islands where there is no water-power, and where it would not be economic to take coal to them. In the Orkney islands, wind-power has been harnessed to experimental generators. New technology is now being used to generate electricity from peat resources as in Ireland, and from tidal and wave-power.

Whatever the method used to produce mechanical energy, the generator produces electrical energy generally at a voltage of 25 000 V. This energy is then fed through the generator switchboards and circuit-breakers to transformers. As their name inplies, the transformers change the voltage of the generators to higher values which are standardised at 132 000 V (132 kV), 275 kV and 400 kV.

The reason for the transformation from a lower voltage to one considerably higher is that it is much more economical to transmit bulk supplies of electrical energy by using the highest voltage possible. In this way, overhead lines and underground cables need have only comparatively small conductors, with the minimum of electrical (I^2R watts) losses. To illustrate this aspect of electricity transmission, a conductor of 18 mm diameter is sufficient to transmit 50 000 kW at 132 000 V. To transmit the same amount of power at 250 V, the conductor diameter would have to be something like 400 mm.

All electricity generated in this country under Statutory requirements is fed into what is known as the 'National Grid'. This is a vast and complex network of overhead lines and underground cables carrying power at high voltages to centres of high load-density. The history of electricity supply is interesting. The first power station was built in Deptford, London, by Ferranti in 1880. He tried to provide a high-voltage ac; he was not successful and this made supply engineers turn their attention to dc (direct current). Distribution was first at 110 V; most of the generation of electricity was in the hands of private companies.

The electricity supply to consumers was rather primitive, though quite workable. One very successful system was devised by Crompton. He stretched bare conductors between glass insulators which were contained in underground ducts or culverts. One such system lasted for some thirty years until, in 1926, it began to give trouble. This method was sufficient for the supply of energy at low voltages, but not satisfactory for high voltages. When it was realised that it was more economical to distribute energy at high voltages, the conductors were given some insulation and were known as cables. One of the earliest attempts to make a high-voltage, paper-insulated cable was by Ferranti. He wrapped oil-impregnated paper round copper rods and pushed the insulated conductors into lengths of iron pipes which he filled with compound.

As the demand for electricity grew, more independent power stations were set up throughout the whole country, either by private companies or by municipal authorities. Eventually the stage was reached where it was recognised that electricity was a real national asset and received the attention of the Government. In 1921 there was the proposal for a national grid system of high-voltage transmission lines which would interconnect all large generating stations. The Electricity Supply Act of 1926 established the Central Electricity Board. This Board was made responsible for the generation of electrical energy in England and Wales. The arrangement in Scotland was separate. In 1948, the whole business of electricity supply was taken over by the British Electricity Authority.

Until recently a body called the Central Electricity Generating Board was responsible for all major power stations and the National Grid and supplied bulk electrical energy to Area Electricity Boards. These Boards were autonomous bodies with a responsibility for the distribution and retail sale of electricity to consumers. The two Scottish Boards owned their power stations and acted as distributors of electricity to their consumers. Essentially the Statutory generation and distribution of electrical energy were a national concern and as such required a body called the Electricity Council to advise the Minister for Energy on all matters affecting the supply industry. The Government, in 1990, began a programme of privatisation which placed the supply of electricity into the control of what are effectively commercial companies with shareholders. Many of these shareholders are now ordinary electricity consumers. Now the supply industry, like other previously nationalised industries, such as steel, railways and gas, has reverted back to the situation before 1948 and it remains to be seen if the new set-up is successful.

The energy at high voltages transmitted by the National Grid system is fed into grid substations for the transformation of the transmission voltage to secondary transmission voltages of 132 kV and 33 kV. This secondary transmission is for large consumers such as factories and those in areas of high load densities. Further substations reduce the secondary transmission voltages to 11 000 V. The term 'distribution' is usually used to refer to the feeding of electrical energy through overhead lines and underground cables to supply small industrial, commercial and domestic premises.

For many years, the mains supply was dc. But gradually as the advantages of ac became apparent, there was a changeover to the latter supply. The voltage of the supply provided by the supply authorities varies according to the consumer and his needs.

Figure 4.1 summarises the pattern of electricity generation, transmission and distribution in Britain today.

If a supply authority agrees to give a supply to a consumer it must comply with the requirements of the Elecricity Supply Regulations. One Clause in particular (No. 34b) is of particular interest. This states that the supply authority must constantly maintain the type of current (dc or ac), the frequency and the declared voltage. The frequency,

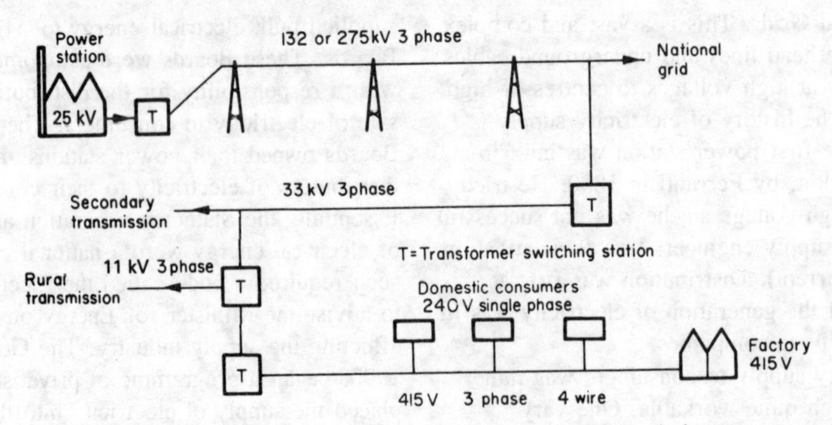

Figure 4.1. The generation, transmission and distribution of electrical energy.

however, may be varied by ± 1 per cent; the declared voltage can be allowed to vary by ± 6 per cent. These percentages are necessary to the supply authorities because variations in loads and voltage drops in cables would otherwise make the supply of electricity at inflexible voltages and frequencies quite uneconomic.

The Electricity Supply Acts define the voltages of supplies:

Extra-low voltage — 30 V (ac) or 50 V dc
Low voltage — 250 V or less
Medium voltage — between 251 V and 650 V
High voltage — between 651 V and 3000 V
Extra-high voltage — exceeding 3000 V (3 kV)

There are two standard systems of ac supply. The first is an arrangement which gives two voltages. This ac system is known as three-phase four-wire and is illustrated in Figure 4.2. The frequency is standard at 50 Hz. There are three 'live' conductors called 'phases' or 'lines'. The voltage between any of these three phases is usually 415 V. If a neutral conductor is earthed, then it will be found that the voltage between any phase conductor and the neutral will be 240 V. The three phases are identified by colours: red, yellow and blue; the neutral is always black. Supplies to premises are always connected to different phases to balance the loads. If the consumer is a small one, a house for instance, the supply cable contains two conductors, a live and neutral; the colour of the live will depend on the phase from which it has been taken. The supply voltage is 240 V, and the system is known as single-phase, two-wire.

Larger consumers receive three-phase, four-wire supplies. The higher voltage is used generally for motors; lighting loads are connected across the outers and the neutral in such a way that when the whole installation is operating, the load across the three phases is reasonably balanced; that is, so that any one phase conductor does not carry a greater current than the other two.

Regulation No. 28a of the 1937 Electricity Supply Act states that energy for use at low voltage must not be introduced by more than one

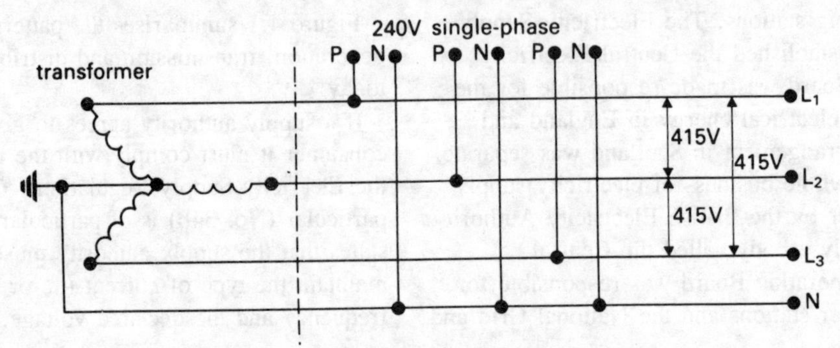

Figure 4.2. The three-phase, four-wire ac system.

pair of conductors of a three-wire or multi-phase system, unless the connected loads exceed 8 kW, *and* it is necessary to do so in order to avoid excessive volt drop or voltage variation on the distribution system in the vicinity. This Regulation does not affect current required for motors or other apparatus which requires medium voltage.

Now a brief word about the conductors used to distribute electrical energy to consumers. First, overhead lines. These are used because transmission by this method is very much cheaper than by underground cables. Indeed, in sparsely populated rural areas, the cost of distribution can be kept reasonably low only by the use of overhead lines. For the National Grid, steel pylons are used. Porcelain and glass insulators carry the conductors which are either copper or aluminium with steel cores for strength. Other materials such as cadmium-copper and phosphor bronze are sometimes used. The choice of the material depends on cost, the required electrical and mechanical properties and on local conditions.

Small lines for rural supply at low and medium voltages use wood or steel poles and have short spans compared with the quite long spans between the Grid pylons. Creosoted wood poles are most common in this country. Insulators are nearly always made from porcelain (with a glazed finish which is self-cleaning). Conductors are generally of stranded copper and uninsulated, but those conductors which run from a pole to a house are insulated.

The type of cable used for supply distribution underground is commonly PVC-sheathed and wire-armoured, with a PVC over-sheath. The conductors are either solid-section aluminium or stranded copper. Paper-insulated cables are also used with stranded copper and aluminium conductors; these cables are required to be made off in sealed cable boxes to prevent the ingress of moisture. They have a higher current rating than PVC types.

The general pattern of low- and medium-voltage distribution in towns and cities is for high-voltage supplies to be taken into a substation, which contains transforming equipment and switchgear. Also included in the circuitry are the protective devices which ensure that faults in cables are detected and prevent damage to equipment and the cables. From these substations are taken cables which supply feeder pillars — these are large boxes so situated that a number of streets can be fed through fuses. The cables are known as feeder cables, from which the service cables (to premises) are connected.

In rural areas, extensive use is made of pole-mounted transformers. The high-voltage input is taken from the main overhead lines. The output conductors from the transformer are sometimes insulated. Where the supply is three-phase, four-wire, there are four conductors. In areas where a good earth is not easily obtainable, a fifth wire, a continuous earth wire (CEW), is sometimes provided by the supply authority.

The sizes of service conductors (the conductors feeding the average domestic or small-consumer installation) is generally 25 mm^2. This is reckoned to be sufficient for a current of about 150 A, enough to cater for the electrical requirements of a house or flat with a floor area not exceeding 100 m^2.

When all the initial details about supply voltage and frequency and number of phases have been settled between the supply authority and the consumer, the form of tariffs must be agreed on. For the normal consumer there are two types of tariff: the ordinary tariff and the off-peak tariff. However, there are many others, depending on the type of consumer.

The domestic tariff applies to premises used excusively as private dwelling houses. Generally a standard charge is levied for each quarter (e.g. £9.37) with each 'unit' (each kilowatt-hour of energy) being charged at typically 6.13p. Most electricity meters are credit meters which are read by the meter-reader twice each year, with the other two quarters being charged on the basis of estimates. At the request of consumers coin (prepayment), token or card-operated meters can be installed, which methods allow for the payment of energy as it is used.

Another tariff offers, in addition to the basic charge, two rates per unit. The first is a very low rate (e.g. 2.24p) per unit (night) supplied during a total of eight hours between 2200 hrs and 0600 hrs. This rate is coupled with a higher rate (e.g. 6.48p) for each (day) unit supplied at all other times.

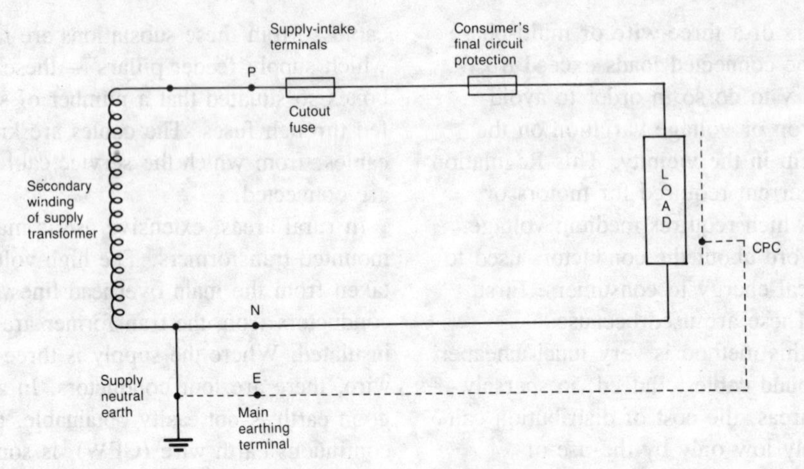

Figure 4.3. Basic diagram for the TNS earthing arrangement.

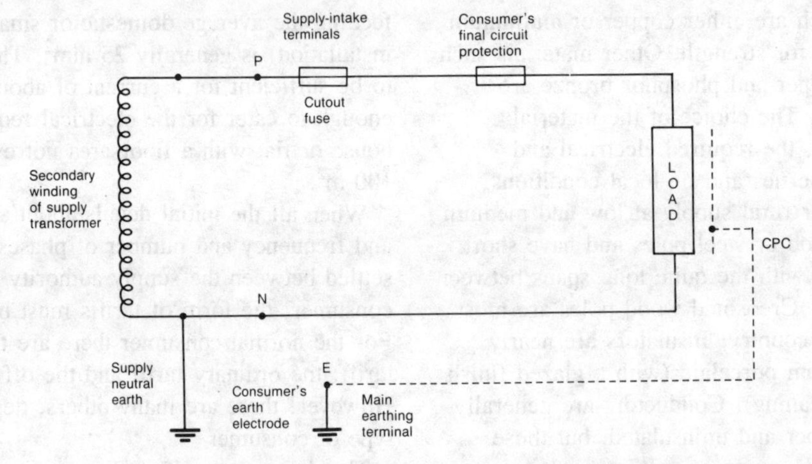

Figure 4.4. Basic diagram for the TT earthing arrangement.

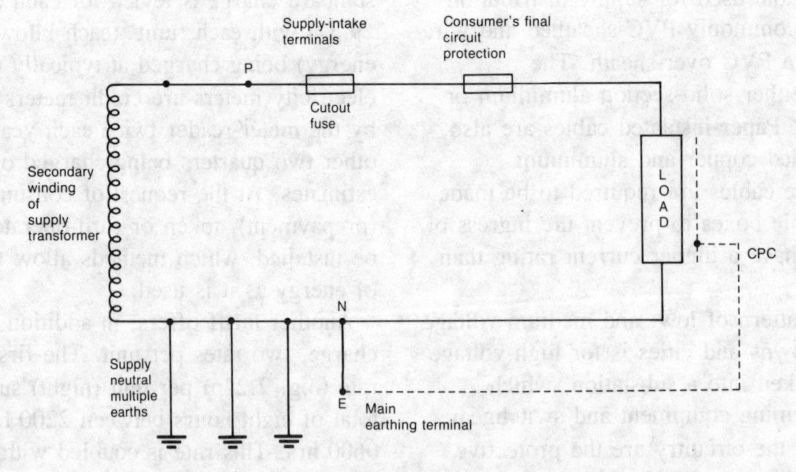

Figure 4.5. Basic diagram for the TNCS earthing arrangement.

Other tariffs are offered to farm consumers whose activities are involved in agriculture, horticulture, floriculture, animal breeding and rearing. Catering tariffs are offered to those establishments involved in the preparation and cooking of food.

Local authorities are offered a public lighting tariff which covers the cost of road and street lighting. Usually this tariff takes in a standing charge for each kilowatt (kW) of load, typically £63.23p per annum, with a charge (e.g. 3.35p) for each unit of energy supplied.

Industrial consumers with large loads are generally charged on a basis of a maximum demand (monthly or quarterly) per kW or kVA (kilovolt-ampere) with an additional charge per unit of energy used. It is often an advantage to the large consumer to take in a supply at a high voltage (e.g. 11 kV or 33 kV) and to provide his own step-down transformer. Most consumers with industrial loads are required to maintain an average power factor of 0.9. If the power factor falls below this figure the charge for maximum demand is increased by a proportionate amount. Most loads made up from electromagnetic equipment and motors tend to require power factor correction to ensure an efficient use of electrical energy. See Chapter 26 'Capacitors and power factor'.

The tariffs or cost for using electrical energy pay for the generation of electricity and its transmission through overhead lines and underground cables. In addition there is the cost of maintaining or replacing such equipment as switchgear and transformers, not to mention the actual cost of the raw fuel, administration and the workforce of the supply industry whose job it is not only to maintain the supply of electricity to consumers but to install new lines, cables and equipment to meet the demand for energy which increases every year.

The electricity supply systems offered in the UK are now identified according to the earthing arrangements as follows:

TNS — This system involves earthing the supply transformer at its star point ('T') with the neutral conductor being separate from the earth return facility ('NS'). The latter is usually the underground supply cable armouring which, on the consumer's side, is connected to the main earthing terminal and at the transformer side is connected to the neutral.

TT — Again the supply transformer star point is earthed ('T') with the consumer providing an earth connection by means of an independent earth electrode ('T'). In such a system, which relies on a good conductive mass of earth for any earth-fault current to flow to its source, a residual-current device is provided to give the consumer added protection against dangerous earth-leakage currents which might arise in the installation should a fault to earth occur.

TNC-S — With the transformer star point earthed ('T') the neutral conductor is earthed at a number of points up to the consumer's supply-intake position where it is then connected to the main earthing terminal. Thereafter, within the consumer's installation, the neutral is separate from the earthing conductors in the installation. This system is sometimes known as a protective multiple earthing (PME) system.

TNC — This system is sometimes known as the earthed concentric wiring system which involves the use of a metal-sheathed conductor (e.g. mineral-insulated cable) which has its sheath acting not only as the neutral or return conductor but also as the circuit protective conductor. The sheath thus has two functions and is known as a PEN (protective earth and neutral) conductor. It can only be used in special situations and stringent requirements must be met before the TNC system is authorised for use in buildings.

These earthing arrangements associated with supply systems in this country are the only ones recognised for the supply of public electricity to consumers. They are intended to offer a greater degree of safety from electic shock. It is thus important that the practising electrician understands the earthing arrangements provided at a client's supply-intake position. If there is doubt he/she should obtain the relevant information from the supply company.

5 Supply-control and distribution on premises

The supply-intake position on any premises, large or small, is generally a matter of agreement between the electrical contractor and the supply-authority's engineer. Some thought is given to the matter because the authority wants the position to be in the most convenient place in the building, so that the cost of the supply equipment (cable, cutouts, and the like) is kept to a minimum. The electrical installation engineer, acting on behalf of his client, is also interested in agreeing on a suitable position so that the electrical installation can be planned with the greatest economy and facility. Not the least important factor, especially where the meter-reader is concerned, is the decision to put the supply-intake position in a place where meters and controls are easily accessible. Too often meters are put in the most awkward positions because if they were put elsewhere they would be unsightly and not acceptable to the householder.

For small installations in towns and cities, the cable is usually two-core, PVC-insulated and provided with sheathing, steel-wire armouring and served with a protective covering of black polyvinyl chloride. The junction which this cable makes with the street-main is contained in a tee-box, generally buried under the pavement just outside the premises. The two-core service-cable conductors are jointed to two of the main cable cores: one to the neutral and the other to one of the phase conductors (red, yellow, or blue). The connectors are either soldered using the usual cable tee-joint or by crimping.

In rural areas, with overhead-line distribution, the house service cables are connected to the line conductors by means of mechanical connectors called line-taps. Conductors to the premises are always insulated, and are in most instances PVC-insulated. The service cables are taken to insulators mounted on D-irons, cleated to the walls of the house, and then run to the supply-intake position.

Whether overhead or underground services, three-phase, four-wire connections are made in a similar manner to the two-wire services.

Whatever the size of the installation, there must be provision for effective control and protection. The IEE regulations indicate what is required to satisfy this requirement. The Regulations make it clear that every consumer's installation must be adequately controlled by specified equipment; and also that this equipment must be readily accessible to the consumer and that it is as near as possible to the supply-intake cutouts. The electricity regulation of the Factories Act says that 'efficient means, suitably located, shall be provided for cutting off all pressure from every part of a system, as may be necessary to prevent danger'.

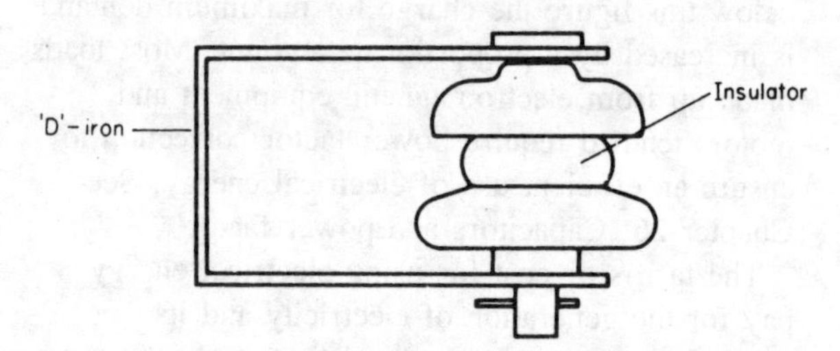

Figure 5.1. D-iron and insulator for overhead rural distribution

The type and size of the main switchgear installed will, of course, depend on the type and the size of the installation and its total maximum load.

Thus, the main switchgear in any installation must be able to:

(*a*) isolate the complete installation from the supply;

(*b*) protect the installation against excess current, which may arise in, say, a short circuit;

(*c*) cut off the current should a serious earth fault occur, say a live conductor touching earthed metalwork.

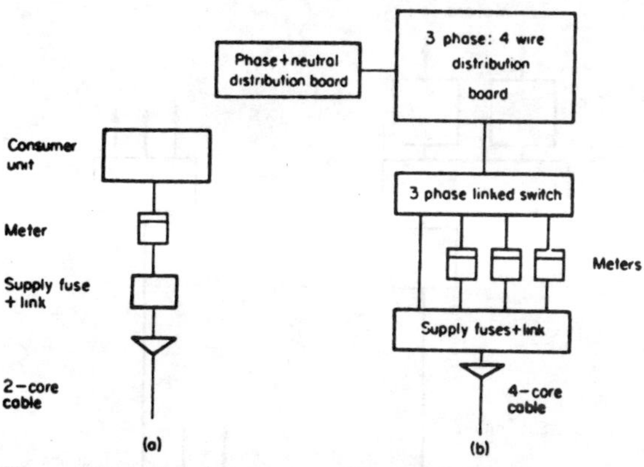

Figure 5.2(*a*). Supply-intake position with consumer unit.

Figure 5.2(*b*). Three-phase, four-wire supply intake position for three- and single-phase supplies.

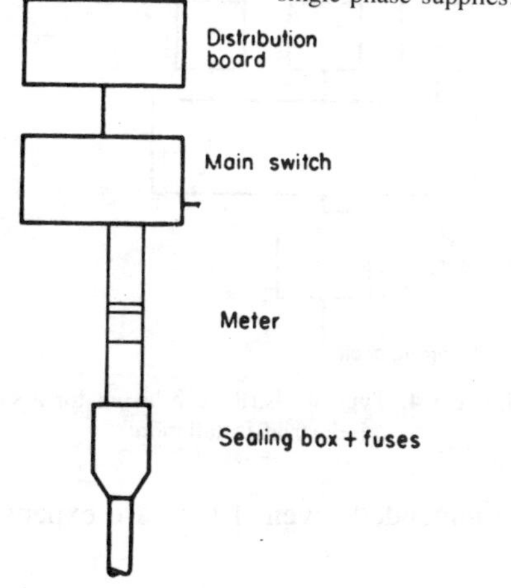

Figure 5.2(*c*). Supply-intake position with main switch and distribution board.

The sequence of supply-control equipment in a single-phase installation is shown in the typical Figure 5.2a. The sequence for a three-phase, four-wire installation is shown typically in Figure 5.2b.

The incoming supply is by underground cable or overhead conductors. If the latter are used, the insulated conductors are taken into the building and terminated in a main fuse cutout and a connector-block for the neutral conductor. Underground cables to most premises are PVC-insulated, armoured and served with PVC. They are terminated in the cutout which also incorporates the connector-block for the neutral. The rating of the service fuse is 80 A for the normal domestic installation, though fuses rated at 100 A are common enough. The higher rating can cope with the loads which are found in houses where there are a number of electrical appliances, storage heating units and the like. The fuse is a BS 88 HBC (high breaking current) type which can cope with current surges while offering the installation full protection should a short-circuit occur. The phase to which the installation is connected (red, yellow or blue) is usually indicated on the cutout.

Access to the cutouts is restricted to all except the supply authority's engineers. To prevent tampering by unauthorised persons, such as unscrupulous householders, the cutouts are sealed, and the seals must remain unbroken.

The next item in the sequence of supply-control equipment is the meter. In installations where a single tariff applies, one meter only is needed. Where, however, two or more tariffs apply, then the metering arrangements must cater for this. The purpose of any meter is to record the amount of electrical energy used by the current-using items connected to the electrical installation. The meter thus records the product of V (volts) and A (amperes) multiplied by t (time). The standard unit is the killowatt-hour (thus 250 V × 4 A × 1 hour = 1 unit or 1 kWh; or 250 V × 2 A × 2 hours = 1 unit or 1 kWh). The meter terminals are sealed against tampering and unauthorised entry.

All the equipment so far described is the property of the supply authority.

From the meter the installation main cables are taken to the main switch or switchfuse, or consumer unit.

The consumer's main switch must be of the double-pole, linked-blade type which will isolate the complete installation from the supply when the switch is operated. If the supply is single-phase, or three-phase and neutral, then all three, or four, poles will be broken.

The main switch is required to isolate the whole of the installation from the supply and, depending on the size of the installation, can be a switch unit or one which contains the consumer's main fuses, known as a switch fuse. In the larger types of installations a moulded-case circuit-breaker is used, which acts not only as a main switch but offers the necessary protection against fault currents.

Distribution boards are used to supply all the

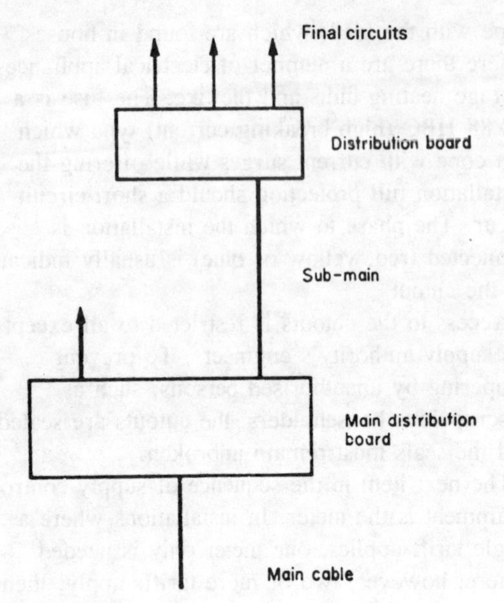

Figure 5.3. Sub-main and final circuit distribution.

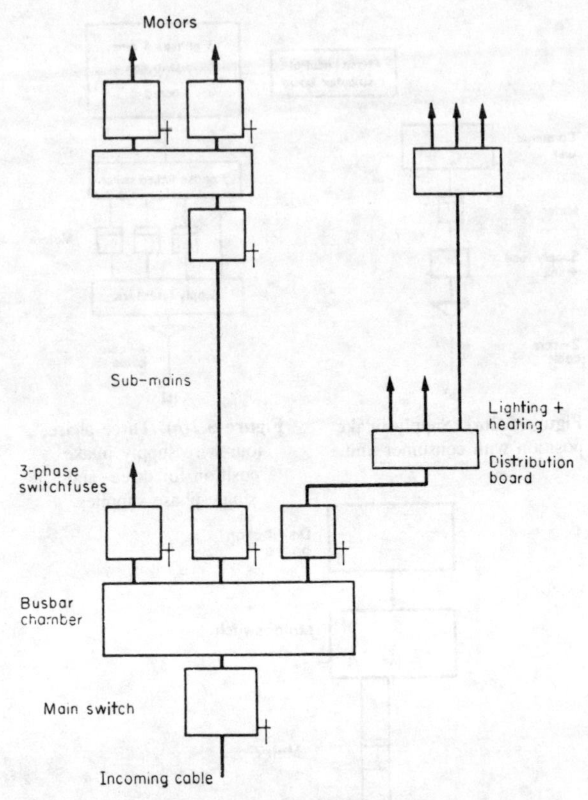

Figure 5.4. Typical distribution layout for a small industrial installation.

final circuits in the installation. They contain the circuit protection devices which are either BS 1361 cartridge fuses or miniature circuit-breakers (MCBs). For certain types of final circuits, BS 88 fuses are needed to give the right kind of protection for, for example, motor loads.

In domestic installations, the distribution board is combined with the main switch and is known as the consumer unit. These vary in capacity, from four-way to twelve-way units. Final circuit protection is by means of BS 3036 fuses (rewirable), BS 1361 fuses (cartridge) or MCBs. Though they are inexpensive, rewirable fuses are not now considered as suitable overcurrent protective devices and most contractors advise clients to install MCBs. These offer good protection and are easy to use by non-electrical persons. They trip off in the event of a fault and indicate the OFF position. If the fault is cleared they can be switched on again. If the fault remains in the circuit they cannot go into the ON position. Fuses, whether of the rewirable or the cartridge type, require replacement by, often, non-electrical persons. Disabled persons may find difficulty in either rewiring a semi-enclosed fuse or even replacing a cartridge fuse. Generally then, the use of MCBs for final circuit protection is always to be recommended, even if they are expensive to install.

The terms used for circuits are important. Main cables are those which carry the total current of the installation. Sub-main cables carry current to sections of a large installation to sub-main switchgear and distribution boards. A final circuit feeds one type of circuit and is not split up to feed another circuit. In a domestic installation a supply is often required for a building which is detached from the main building, such as a garage. In this case a final circuit in the consumer unit feeds a cable taken into the garage which must be terminated in, say, a four-way consumer unit from which the garage lighting and socket-outlets are taken. It is a requirement of the Regulations that every detached building is provided with its own means of isolation.

Industrial installations, with a three-phase, four-wire supply, often have rather more complicated arrangements at their supply-intake positions.

In multi-storeyed buildings, the supply (three-phase, four-wire) is taken into a basement room. Then copper conductors of large cross-section are installed to form what are called 'rising mains'. Each flat or floor is tapped off these conductors. The arrangement of the supply-intake position for each consumer is then much the same as found in ordinary domestic premises.

All switchgear and control gear for any installation must be of suitable capacity. This means that the gear must be capable of carrying, without damage to itself or its associated wiring, the current that will flow when the installation is being operated in normal circumstances. The Regulations require that 'diversity factor' may be applied when working out the size of conductors for main and submain cables, provided the conditions of operation are known.

The diversity factor is used to obtain a realistic indication of the current which would flow in normal use. This can be seen in the lighting provision of an installation where, at any one time, only a small part of the installation is being used; a smaller current thus flows. The diversity factor is applied only to main cables and main switchgear which feed final circuits, but not to the final circuits themselves.

A final circuit can range from a pair of 1.5 mm^2 cables feeding a light to a very heavy three-core cable feeding a large motor from a circuit-breaker or switch at a main switchboard. The one important rule which applies to final circuits is that mentioned in the Electricity Supply Regulation 27: 'All conductors and apparatus must be of sufficient size and power for the work they are called on to do, and so constructed, installed and protected as to prevent danger.' The IEE Regulations ensure that this Regulation is complied with.

The IEE Regulations deal with the requirements for disconnecting circuits from the supply in the event of overload and short circuits. The rating of a protective device (fuse or circuit-breaker) must not be less than the designed load current of the circuit and, also, that rating should not exceed the current-carrying capacity of the lowest-rated conductor in a circuit.

There are five important general groups of final circuits:

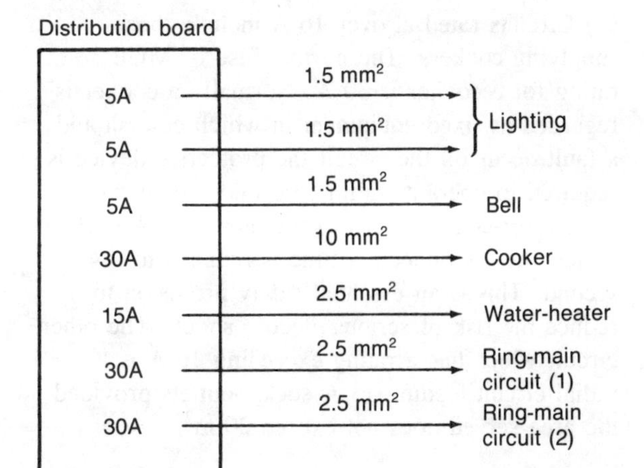

Figure 5.5. Typical final circuits from a consumer unit.

(*a*) Rating not exceeding 16 A.
(*b*) Rating exceeding 16 A.
(*c*) Rated for 13 A fused plugs.
(*d*) Rated for feeding fluorescent and other discharge-lamp circuits.
(*e*) Rated for feeding a motor.

(*a*) Final circuits which have a rating not exceeding 16 A include lighting circuits and those supplying 3 kW water-heaters. Domestic lighting circuits are usually rated at 5 A (fuse or MCB). Commercial or industrial lighting circuits may be protected by overcurrent devices rated at 6 A or 16 A and this applies whether the type of lamp-holder is bayonet cap or Edison screw. Domestic water-heating circuits are protected by a 15 A fuse or MCB.

It is recommended that all domestic installations have at least two lighting circuits. The reason is simply to ensure that should one lighting circuit protection device operate, the other circuit will still provide some lighting. There is also a general recommendation that no lighting circuit should have more than ten lighting points. The assumed current demand for a lighting outlet is the current equivalent of 100 W. Therefore for ten 100 W lamps, the circuit current, should all the lamps be switched on, would be just over 4 A, which level of current approaches the current rating of a 5 A fuse or MCB.

(*b*) Circuits rated at over 16 A include those supplying cookers. The normal fuse or MCB rating for a cooker is 30 A. Normally a cooker is regarded as fixed equipment in which case should a fault occur on the circuit the protective device is required to operate within 5 seconds. If, however, the cooker-control unit incorporates a 13 A socket-outlet, the disconnection time is reduced to 0.4 second. This is an essential safety provision to reduce the risk of serious electric shock. The other circuit which has a rating exceeding 16 A is a radial circuit feeding 13 A socket-outlets provided the area served does not exceed 20 m^2.

(*c*) The first suggestion of a standard socket-outlet for domestic purposes and provided with a fused plug was made in 1944, in connection with a report on electrical installations in postwar buildings.

Up until the outbreak of the Second World War, the range of types of socket-outlets and plugs was so great that if one went from one part of the country to another, one had to replace all one's old plugs or appliances. As a solution to this problem, the standard socket-outlet and plug came into existence and is now accepted by virtually all installation engineers. Basically, the 13 A circuits use socket-outlets with rectangular sockets. The appropriate plug has rectangular pins and contains a cartridge fuse the rating of which depends on the amount of current taken by the appliance connected to the plug (3 A or 13 A).

The Wiring Regulations deal with circuits for socket-outlets fitted with fused plugs and rated not more than 13 A. The IEE Regulations indicate how these socket-outlets are to be connected. It will be noticed that a diversity factor has been applied. For instance, two socket-outlets can be wired with 2.5 mm^2 cables and protected by a 20 A fuse. The reason for this seeming contravention of the Regulations is that it is unlikely that both 13 A socket-outlets would ever supply their full rated current. This is even more the case in a ring-main circuit where some socket-outlets never supply more than is needed for a table lamp, a radio or television set (i.e. less than 1.5 A).

Figure 5.6 shows a typical ring-circuit

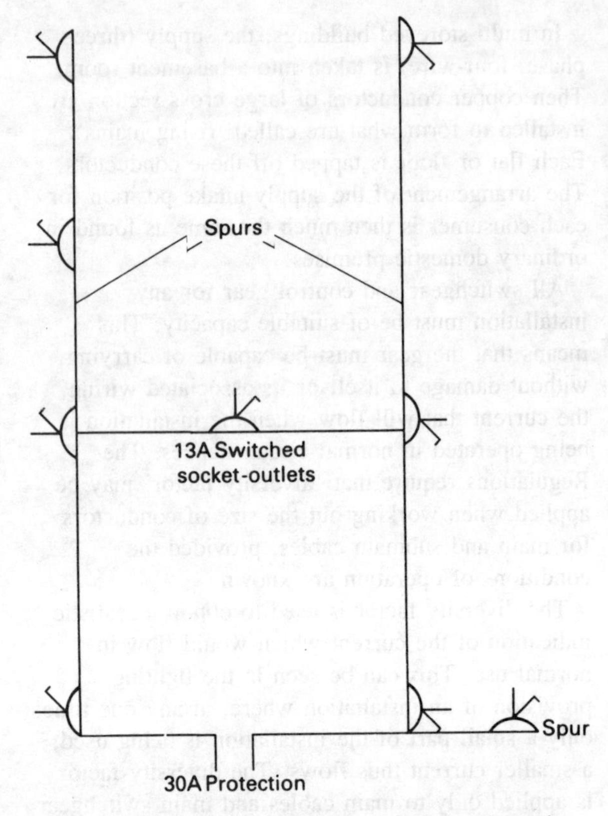

Figure 5.6. Typical ring-main circuit with 13 A socket-outlets.

arrangement which complies fully with the Regulations' requirements. Thirteen-amp socket-outlets with fused plugs should be used only on ac systems. The maximum number of 13 A socket-outlets connected in any one circuit will depend on the estimated peak loading of the circuit. The total permissible load of fixed appliances fed from a single final circuit must not exceed 16 A.

(*d*) One of the main requirements so far as the electrician is concerned with regard to fluorescent lighting circuits is the switching. In such circuits, it is not sufficient to take the total wattage of the lamps (as one would do with filament lamps, taking a minimum of 100 W for any lampholder). Other factors must be considered. The inclusion of a choke or inductor increases the current taken by the circuit. The Regulations require that circuits shall be capable of carrying the total steady current, viz. that of the lamps and any associated gear, that is, lamp watts multiplied by not less than 1.8. This factor is based on the assumption that a circuit is corrected to a power factor of not less than 0.85. and takes into account control gear losses.

In the early days when fluorescent lighting was beginning to become popular, much trouble was experienced with the tumbler switches normally used for filament-lighting circuits. When they were switched off they often broke down between the live side and earth. This was due to the momentary high voltage induced by breaking an inductive circuit. The Regulations require that any switch controlling circuits comprising discharge lamps should have a rating of not less than twice the steady current in the circuit it controls, unless the switch is specially designed to break an inductive load at its full rated capacity. The recommended switch-type is quick-make/slow-break, or else ac rated.

(*e*) Final circuits which feed motors require some special consideration, though for the most part they must comply with the Regulations which apply to other types of final circuits. The current rating of cables feeding a motor should be based on the full-load current taken by the motor. More than one motor may be connected to a 16 A final circuit, provided that the total full-load rating of the motors does not exceed the rating of the smallest cable in the circuit. If, however, the rating of the motor does exceed 16 A, then the circuit must supply one motor only.

If any conductor carries a current, it will 'drop' voltage: from Ohm's law, $V = IR$, where I is the current flowing in the conductor and R is the conductor resistance. The Wiring Regulations allow a maximum of 4 per cent drop between the origin of a circuit and its load terminals. This means that on 240 V nominal voltage, the loss in voltage can be 9.6 V, giving the voltage at the load terminals of 230.4 V. The *nominal* voltage is the voltage offered by the supply authority but is not necessarily the actual voltage at the supply-intake position. This in fact is allowed by law to vary by ± 6 per cent. The Wiring Regulations, however, recognise this might create a problem for certain types of electrical equipment which might require its rated voltage in order to function properly and safely. In the worst situation, if the supply voltage falls to its lower −6 per cent limit, this, coupled with the maximum voltage drop allowed on a final circuit, would mean a possible loss of up to 10 per cent of the voltage at the terminals of a load.

Certain loads such as fluorescent lighting are not greatly affected by voltage loss. However, filament lamps and heating elements are seriously affected by a loss of voltage. Taking a 1 kW heating element as an example, at its rated voltage of, say, 240 V it will deliver 1000 W of heat. If the voltage falls by 4 per cent, the heat output will decrease to 920 W, which is around 8 per cent loss. In a similar situation the light output of a filament lamp would also fall. In general, the designer of an installation will have to make a decision on the voltage drop allowed on any final circuit and if the voltage requirement is critical, may have to specify a conductor of increased cross-sectional area (csa) to ensure that the voltage at the load terminals does not fall below that which renders the load inefficient in electrical terms.

6 Conductors and cables

A 'conductor' in electrical work means a material which will allow the free passage of an electric current along it and which presents very little resistance to the current. If the conducting material has an extremely low resistance (for instance a copper conductor) there will be only a slight warming effect when the conductor carries a current. If the conductor material has a significant resistance (for instance, iron wire) then the conductor will show the effects of the electric current passing through it, usually in the form of an appreciable rise in temperature to produce a heating effect.

A 'cable' is defined as a length of insulated conductor (solid or stranded), or of two or more such conductors, each provided with its own insulation, which are laid up together. The conductor, so far as a cable is concerned, is the conducting portion, consisting of a single wire or of a group of wires in contact with each other.

The practical electrician will meet two common conductor materials extensively in his work: copper and aluminium.

As a conductor of electricity, copper has been used since the early days of the electrical industry because it has so many good properties. It can cope with onerous conditions. It has a high resistance to atmospheric corrosion. It can be jointed without any special provision to prevent electrolytic action. It is tough, slow to tarnish, and is easily worked. For purposes of electrical conductivity, copper is made with a very high degree of purity (at least 99.9 per cent). In this condition it is only slightly inferior to silver.

Aluminium is now being used in cables at an increasing rate. Although reduced cost is the main incentive to use aluminium in most applications, certain other advantages are claimed for this metal. For instance, because aluminium is pliable, it has been used in solid-core cables. Aluminium was used as a conductor material for overhead lines about seventy years ago, and in an insulated form for buried cables at the turn of the century. The popularity of aluminium increased rapidly just after the Second World War, and has now a definite place in electrical work of all kinds.

The following table shows the properties of copper compared with aluminium:

	Aluminium	*Copper*
Density at 20°C (kg/cm^3)	2700	8890
Melting point (°C)	659	1083
Electrical conductivity at 20°C (per microhm metre)	62	100
Resistivity at 20°C (in microhm metre)	0.0283	0.0174

In the above table, copper is taken as annealed high conductivity and aluminium as having a 99.5 per cent purity.

Aluminium has an excellent resistance to corrosion, though simple precautions must be taken to avoid corrosion in particular situations.

Conductors

Conductors as found in electrical work are most commonly in the form of wire or bars and rods. There are other variations, of course, such as machined sections for particular electrical devices (e.g. contactor contacts). Generally, wire has a flexible property and is used in cables. Bars and rods, being more rigid, are used as busbars and earth electrodes. In special form, aluminium is used for solid-core cables.

Wire for electrical cables is made from wire-bars. Each bar is heated and passed through a series of grooved rollers until it finally emerges in the form of a round rod. The rod is then passed through a series of lubricated dies until the final diameter of wire is obtained. Wires of the sizes generally used for cables are hard in temper when drawn and so are annealed at various stages during the transition from wire-bar to small-diameter wire. Annealing involves placing coils of the wire in furnaces for a period until the metal becomes soft or ductile again.

Copper wires are often tinned. This process was first used in order to prevent the deterioration of the rubber insulation used on the early cables. Tin is normally applied by passing the copper wire through a bath containing molten tin. With the increasing use of plastics materials for cable insulation there was a tendency to use untinned wires. But now many manufacturers tin the wires as an aid in soldering operations. Untinned copper wires are, however, quite common. Aluminium wires need no further process after the final drawing and annealing.

All copper cables and some aluminium cables have conductors which are made up from a number of wires. These conductors are of two basic types: stranded and bunched. The latter type is used mainly for the smaller sizes of flexible cable and cord. The solid-core conductor (in the small sizes) is merely one single wire.

Most stranded conductors are built up on a single central conductor. Surrounding this conductor are layers of wires in a numerical progression of 6 in the first layer, 12 in the second layer, 18 in the third layer and so on. The number of wires contained in most common conductors are to be found in the progression 7, 19, 37, 61, 127.

Stranded conductors containing more than one layer of wires are made in such a way that the direction of lay of the wires in each layer is of the reverse hand to those of adjacent layers. The flexibility of these layered conductors is good in the smaller sizes (e.g. 7/0.85 mm) but poor in the larger sizes (e.g. 61/2.25 mm).

When the maximum amount of flexibility is required, the 'bunching' method is used. The essential difference of this method from 'stranding' is that all the wires forming the conductor are given the same direction of lay. A further improvement in flexibility is obtained by the use of small-diameter wires, instead of the heavier gauges as used in stranded cables.

When more than one core is to be enclosed within a single sheath, oval and sector-shaped conductors are often used. These shaped conductors are shown in Figure 6.1.

It is of interest to note that when working out the dc resistance of stranded conductors, allowance must be made for the fact that, apart from the central wire, the individual strands in a stranded conductor follow a helical path — and so are slightly longer than the cable itself. The average figure is 2 per cent. This means that if a stranded conductor is 100 m long, only the centre strand is this length. The other wires surrounding it will be anything up to 106 m in length.

Because aluminium is very malleable, many of the heavier cables using this material as the conductor have solid cores, rather than stranded. A saving in cost is claimed for the solid-core aluminium conductor cable.

Conductors for overhead lines are often strengthened by a central steel core which takes the weight of the copper conductors between the

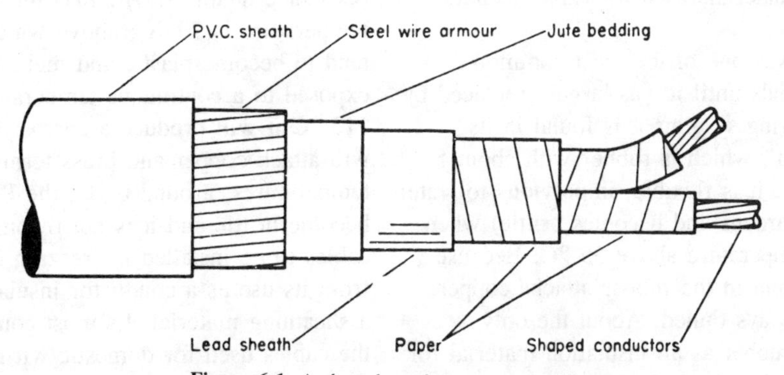

Figure 6.1. A shaped-conductor power cable.

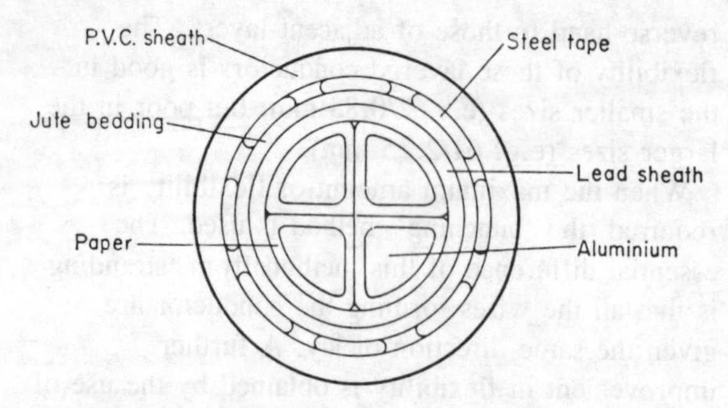

Figure 6.2. Section of an aluminium-core power cable.

poles or pylons. Copper and aluminium are used for overhead lines.

Conductor sizes are indicated by their cross-sectional area (csa). Smaller sizes tend to be single strand conductors; larger sizes are stranded. Cable sizes are standardised, starting at 1 mm^2, then increasing to 1.5, 2.5, 4, 6, 10, 16, 25 and 35 mm^2. As cable sizes increase in csa the gaps between them also increase. The large sizes of armoured mains cable from 25 mm^2 tend to have shaped stranded conductors.

Insulators

Many materials are used for the insulation of cable conductors. The basic function of any cable insulation is to confine the electric current to a definite path; that is, to the conductor only. Thus, insulating materials chosen for this duty must be efficient and able to withstand the stress of the working voltage of the supply system to which the cable is connected. The following are some of the more common materials used for cable insulation:

Rubber. This was one of the most common insulating materials until it was largely replaced by PVC. In old wiring systems it is found in its 'vulcanised form', which is rubber with about 5 per cent sulphur. It is flexible, impervious to water but suffers (it hardens and becomes brittle) when exposed to a temperature above 55 °C. Because the sulphur content in the rubber attacks copper, the wires are always tinned. About the only application for rubber as an insulation material for conductors nowadays is in domestic flexibles used for hand appliances such as electric irons. The working temperature is 60 °C.

85 °C Rubber. This material is a synthetic rubber designed for working temperatures up to 85 °C. It is in its flexible cord format used for hot situations such as immersion heaters and night storage heaters where the heat from elements can travel into the flexible conductors. As a sheathing material it is susceptible to oil and grease and thus such flexibles are sheathed with chloro-sulphonated polyethylene (CSP). This type of sheath is known as HOFR (heat and oil resisting and flame retardant). Often used for heavy-duty applications, it is found in its larger csa sizes feeding exterior equipment such as mobile cranes and conveyors.

Silicone rubber. This material is sometimes designated 150 °C insulation and can operate in a continuous temperatrure up to that level. Applications of this fire-resistant cable include the wiring of fire alarm, security and emergency lighting circuits where there is a need for these circuits to function in fire condititons. It is also useful when connections have to be made to terminals in enclosures in which heat might be considerable, such as in enclosed lamp fittings and heaters.

PVC. This material is polyvinyl chloride and is now the most common insulating material used for cables and flexibles at low voltages. Its insulating properties are actually less than those for rubber. However it is impervious to water and oil and can be self-coloured without impairing its insulation resistance qualities. The maximum working temperature is 70 °C, above which the PVC will tend to become plastic and melt. If PVC is exposed to a continuous temperature of around 115 °C it will produce a corrosive substance which will attack copper and brass terminals. At low temperatures, around 0 °C, the PVC tends to become brittle and it is not recommended for PVC cables to be installed in freezing conditions. Apart from its use as a conductor insulation, it is used as a sheathing material. Its most common form is in the cables used for domestic wiring and for domestic flexibles.

Paper. Paper has been used as an insulating material from the very early days of the electrical industry. The paper, however, is impregnated to increase its insulating qualities and to prevent its being impaired by moisture. Paper-insulated cables, usually of the large csa sizes, are terminated in cable boxes sealed with resin, or compound, to prevent the ingress of moisture. The cables are sheathed with lead and armoured with steel or aluminium wire or tape. Such cables are mainly used for large loads at high voltages.

Mineral insulation. This is composed of magnesium oxide powder and is used in the type of cable known as MIMS (mineral-insulated metal-sheathed) with the sheath usually made from copper. It was originally developed to withstand both fire and explosion, but is now used for more general applications. The cable is non-ageing and can be operated with sheath temperatures of up to 250 °C. Because the magnesium oxide is hygroscopic (it absorbs moisture) the cable ends must always be sealed. The temperature limits of the seals depend on the cable's application.

Glass insulation. This material is very heat-resistant and is used for temperatures as high as 180 °C. As glass-fibre, the insulation takes the form of impregnated glass-fibre lappings, with impregnated glass-fibre braiding. This insulation is found commonly in the internal wiring of electric cookers or other appliances where the cable must be impervious to moisture, resistant to heat and be tough and flexible.

Protection

Sheathing. Only in exceptional circumstances does the insulation of a conductor offer some protection against attacks by water, oils, acids and mechanical damage. Thus, it is common practice to protect the insulated conductor by a sheath or covering of some material which will enable the cable to be used in situations where some physical damage might result.

The basic purpose of the sheath is to prevent moisture from reaching the insulated core of the cable when in service. This implies that the sheath be impervious and resistant to corrosion. Once applied, a sheath must be sufficiently pliable to withstand a number of coiling and straightening operations during cable installation. Sheathing materials vary considerably, and are usually associated with the type of material used for the conductor installation. PVC-insulated conductors are sheathed with the same material. Mineral-insulated conductors are enclosed within a metal sheath which can be copper (MICS) or aluminium (MIAS). Paper-insulated cables generally have a lead-alloy sheath. Aluminium conductors are used with aluminium sheaths.

In many instances, the metal sheathing and armouring of cables are used to act as a conductor for earth-leakage currents.

Sometimes the wiring system acts as a sheath to protect against damage to the cables. For instance, conduit protects PVC-insulated cables and the cables need not be provided with a sheath.

Armouring. In certain circumstances it is necessary for a cable to be protected against the occurrence of mechanical damage. Protection by 'armouring' is defined as the provision of a 'helical' wrapping or wrappings of metal (usually wires or tapes), primarily for the purpose of mechanical protection. The type of damage against which the cable is protected is rough treatment, abrasion, collision. The materials used, in tape or wire form, for armouring cables is most often steel. But aluminium is also used.

Special cables and conductors

There are many applications for cables and conductors apart from the most common use, which is wiring to form part of an installation for lighting and current-using apparatus. Some of the most common types will be mentioned here, but the practical electrician may come across many other applications and types.

Possibly the most familiar conductor is that used for extra-low voltage work, for bells and other similar signalling applications. The feature of this conductor is the relatively thin covering of insulation — because the circuit would never be supplied with anything greater than extra-low voltage (see IEE Regulations Definition: 50 V ac or 120 V dc). The current carried is small.

Another type of conductor is called 'fittings wire', which is suitable only for the internal wiring of fittings with small bore aperture and not subject to disturbance or mechanical damage. The current carried is quite small.

Also of small csa are the telephone cables used not only for GPO work but also for the internal communication systems in offices and works; these systems are often installed by a contracting electrician with experience in this particular field.

Winding wires are yet another type of conductor used for a particular application. In this instance they are used for making the coils of electromagnets and solenoids. The conductor material is nearly always high-conductivity copper with various insulating materials applied. The type of insulation depends on the application of the coil, and is generally one of three groups: paper, textiles and enamels. For transformer windings paper is used. In the textile group fall cotton, silk (natural and synthetic) and glass-fibres. Various types of varnished cloth are also to be found. In the enamelled-wire field there are types of enamels ranging from ordinary applications to those involving working temperatures of above 180 °C.

The following are a few of the applications for winding wires formed into coils of one sort or another: small transformers, chokes, relays, solenoids; motor and transformer windings; electronic devices.

Soil-warming conductors are generally supplied with extra-low voltage and are bare, specially selected, galvanised steel wires. The gauge is from 2.6 to 1.6 mm, the larger gauges being more resistant to corrosion and mechanical damage. For small domestic soil-warming applications, the circuit is fed from a double-wound transformer supplying 4 to 8 V. For larger installations the voltage is up to 30 V. The wires are really resistive conductors which dissipate heat when a current passes through them.

The elements of electrical heating apparatus are yet another group of conductors, possessing sufficient resistance to produce an effect: heat. Most element conductors are made from nickel-chromium, or nickel-iron-chromium, though other alloys are used depending on the application, and usually the final temperature required.

One now-common application of the resistive conductor or element is for surface heating; when provided with some suitable insulation they become heating cables. The applications include maintaining an easy-flow temperature of liquids in pipes, for heating jackets and soil-warming. An example of the latter is a central copper conductor, PVC-insulated, spiralled with a resistance heating element. A braiding of tinned-copper wire is provided and the whole is sheathed with PVC. These cables are designed for use on normal mains voltage (200-250 V) to eliminate the need for transformers for reducing the mains voltage to 8-30 V. The cables are manufactured to give so many watts per metre length.

Cable types

The range of types of cables used in electrical work is very wide: from heavy lead-sheathed and armoured paper-insulated cables to the domestic flexible cable used to connect a hair-drier to the supply. Lead, tough-rubber, PVC and other types of sheathed cables used for domestic and industrial wiring are generally placed under the heading of power cables. There are, however, other insulated copper conductors (they are sometimes aluminium) which, though by definition are termed cables, are sometimes not regarded as such. Into this category fall those rubber and PVC insulated conductors drawn into some form of conduit or trunking for domestic and factory wiring, and similar conductors employed for the wiring of electrical equipment. In addition, there are the various types of insulated flexible conductors including those used for portable appliances and pendant fittings.

The main group of cables is 'flexible cables', so termed to indicate that they consist of one or more cores, each containing a group of wires, the diameters of the wires and the construction of the cables being such that they afford flexibility.

Single-core. These are natural or tinned copper wires. The insulating materials include butyl-rubber (known also as 85 °C rubber-insulated cables), silicone-rubber (150 °C, EP-rubber (ethylene propylene), and the more familiar PVC.

The synthetic rubbers are provided with braiding and are self-coloured. The IEE Regulations recognise these insulating materials for twin- and multi-core flexible cables rather than for use as single conductors in conduit or trunking wiring systems. but they are available from cable manufacturers for specific installation requirements. Sizes vary from 1.00 to 36 mm^2 (PVC) and 50 mm^2 (synthetic rubbers).

Two-core. Two-core or 'twin' cables are flat or circular. The insulation and sheathing materials are those used for single-core cables. The circular cables require cotton filler threads to gain the circular shape. Flat cables have their two cores laid side by side.

Three-core. These cables are the same in all respects to single- and two-core cables except, of course, they carry three cores.

Composite cables. Composite cables are those which, in addition to carrying the current-carrying circuit conductors, also contain a circuit-protective conductor.

To summarise, the following groups of cable types and applications are to be found in electrical work, and the electrician, at one time or another during his career, may be asked to install them.

Wiring cables. Switchboard wiring; domestic and workshop flexible cables and cords. Mainly copper conductors.

Power cables. Heavy cables, generally lead-sheathed and armoured; control cables for electrical equipment. Both copper and aluminium conductors.

Mining cables. In this field cables are used for trailing cables to supply equipment; shot-firing cables; roadway lighting; lift-shaft wiring; signalling, telephone and control cables. Adequate protection and fireproofing are features of cables for this application field.

Ship-wiring cables. These cables are generally lead-sheathed and armoured, and mineral-insulated, metal-sheathed. Cables must comply with Lloyd's Rules and Regulations, and with Admiralty requirements.

Overhead cables. Bare, lightly-insulated and insulated conductors of copper, copper-cadmium and aluminium generally. Sometimes with steel core for added strength. For overhead distribution cables are PVC and in most cases comply with British Telecom requirements.

Communications cables. This group includes television down-leads and radio-relay cables; radio-frequency cables; telephone cables.

Welding cables. These are flexible cables and heavy cords with either copper or aluminium conductors.

Electric-sign cables. PVC- and rubber-insulated cables for high-voltage discharge lamps (neons, etc.) able to withstand the high voltages.

Equipment wires. Special wires for use with instruments, often insulated with special materials such as silicone, rubber and irradiated polythene.

Appliance-wiring cables. This group includes high-temperature cables for electric radiators, cookers and so on. Insulation used includes nylon, asbestos and varnished cambric.

Heating cables. Cables for floor-warming, road-heating, soil-warming, ceiling-heating and similar applications.

Flexible cords. A flexible cord is defined as a flexible cable in which the csa of each conductor does not exceed 4 mm^2. The most common types of flexible cords are used in domestic and light industrial work. The diameter of each strand or wire varies from 0.21 to 0.31 mm. Flexible cords come in many sizes and types; for convenience they are grouped as follows:

Twin-twisted. These consist of two single insulated stranded conductors twisted together to form a two-core cable. Insulation used is vulcanised rubber and PVC. Colour identification in red and black is often provided. The rubber is protected by a braiding of cotton, glazed-cotton, rayon-braiding and artificial silk. The PVC-insulated conductors are not provided with additional protection.

Three-core (twisted). Generally as twin-twisted cords but with a third conductor coloured green, for earthing lighting fittings.

Twin-circular. This flexible cord consists of two conductors twisted together with cotton filler-threads, coloured brown and blue, and enclosed within a protective braiding of cotton or nylon. For industrial applications, the protection is tough rubber or PVC.

Three-core (circular). Generally as twin-core circular except that the third conductor is coloured green and yellow for earthing purposes.

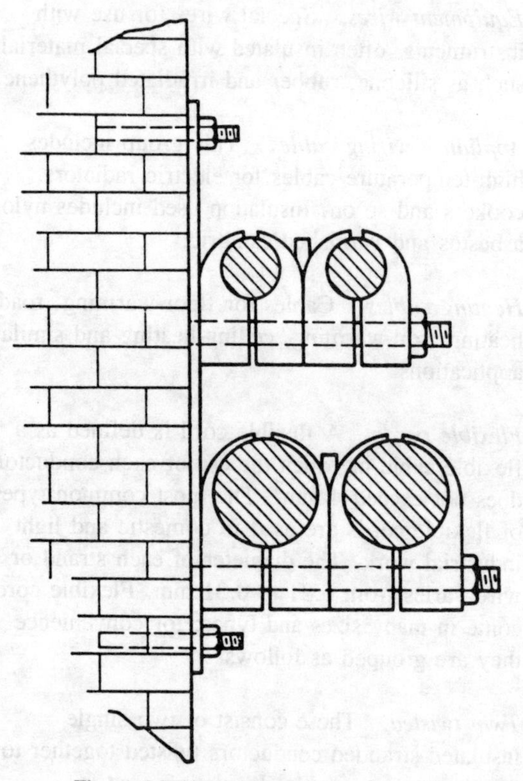

Figure 6.3. Cables supported by claw cleats.

Four-core (circular). Generally as twin-core circular. Colours are brown and blue.

Parallel-twin. These are two stranded conductors laid together in parallel and insulated to form a uniform cable with rubber or PVC

Twin-core (flat). This consists of two stranded conductors insulated with rubber, coloured red and black, laid side by side and braided with artificial silk.

High-temperature lighting, flexible cord. With the increasing use of filament lamps which produce very high temperatures, the temperature at the terminals of a lampholder can reach 71 °C or more. In most instances the usual flexible insulants (rubber and PVC) are quite unsuitable and special flexible cords for lighting are now available. Conductors are generally of nickel-plated copper wires, each conductor being provided with two lappings of glass-fibre. The braiding is also varnished with silicone. Cords are made in the twisted form (two- and three-core).

Flexible cables

These cables are made with stranded conductors, the diameters being 0.3, 0.4, 0.5 and 0.6 mm. They are generally used for trailing cables and similar applications where heavy currents up to 630 A are to be carried, for instance, to welding plant.

Cable installation

The installation of cables is dealt with fully in Chapter 11, 'Installation methods'. However, one or two relevant points can be made here, particularly with regard to cable supports, the identification of conductors, and cable bends.

The Wiring Regulations give guidance on how the supports for electric cables are regulated with regard to the distance between them. In general, the heavier the cable, then the more support there should be per unit length (metre). The spacing is different for vertical and horizontal cable runs, the shorter spacing being for the horizontal runs. Cable supports include clips made from metal and in the form of buckles, or cleats, generally porcelain for smaller types of cables to claw-type cleats for the larger cables. Figures 6.3 and 6.4 show a variety of cable supports. Saddles are also used to hold single-core, two- and multi-core cables, and a group of single-core cables.

It is important to ensure that when a cable is bent the radius of the bend is not too small; that is the bend should not be sharp. The IEE Regulations indicate the minimum internal radius of bends for cables of different types, and consider the type of insulation, the covering (sheath and armouring) and the diameter.

Conductor identification

The Wiring Regulations require that all conductors have to be identified by some means to indicate their function. For example, the phase conductors of a three-phase system are coloured red, yellow and blue, with the neutral coloured black. Protective conductors are identified by green/yellow.

Note: Colour is also used to identify certain types of circuits. For example, if MI cable is used for fire alarm circuits, a PVC oversheath is provided, coloured red. If MI cable is used for emergency lighting, the PVC oversheath is white. Electrical services in a building, such as a factory where pipes carrying gas, oil and hot water are installed, are coloured a light orange. In general, conductor identification methods are as follows:

1. Colouring of the conductor insulation.
2. Printed numbers on the conductor insulation (e.g. on paper).
3. Coloured adhesive tapes at the terminations of the conductors.
4. Coloured sleeves at the terminations of the conductors.
5. Numbered sleeves at the terminations of the conductors.
6. Coloured paint for bare conductors (busbars and the like).
7. Coloured discs fixed near to the terminations of conductors (e.g. on a meter board).

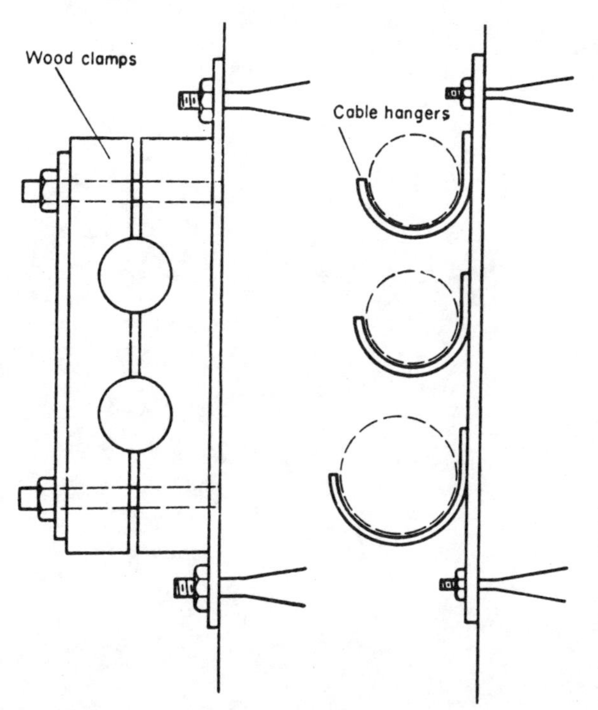

Figure 6.4. Cable supports: wood cleats and hangers.

As a point of interest, the British Standards Institution require that the colouring for cable identification should be fast, that is that it should not fade on prolonged exposure to light. For instance, PVC colours tend to fade in sunlight. The printing of numbers and letters must be such that they cannot be rubbed off in use.

General. One problem with PVC insulation is that it produces toxic fumes when it burns and the electrician is advised to keep clear of such situations. Developments in recent years have resulted in a low-toxicity PVC, known as PX and FP200, which contributes a significant degree of safety for those who have to deal with fires. PX cables have low-toxicity PVC insulation and sheathing with an aluminium foil for added protection. FP200 cables have silicone rubber PVC insulation, and are aluminium sheathed.

Old wiring systems

While this chapter has dealt with all the many wiring systems used in electrical installations over the years, some have been replaced by new systems. For example, VRI systems will now only be found in older installations; and lead-sheathed cables are not commonly available, being largely replaced by mineral-insulated copper-sheathed cables. However, these systems will still be found by electricians engaged in, say, the rewiring of houses and other premises. If a complete rewire is undertaken, then the old system is removed completely. Problems arise when an extension or alteration to an existing system is required by a client. While the old system might still be electrically sound, in terms of insulation resistance, there is the need to advise the client that the older part of the installation should be updated to comply with the requirements of the current edition of the Wiring Regulations, particularly in respect of protection against electric shock and overcurrent.

The hard fact about old installations is that they are now contributing to a dramatic increase in the occurrence of building fires due to electrical causes. Often neglected and rarely inspected and tested over many years, they now show their age. The Electrical Contractors Association in England, Wales and Scotland mount regular advertising campaigns to warn householders of the dangers inherent in old wiring systems.

7 Conductor joints and terminations

Basic electrical and mechanical requirements

The following are the basic requirements which must be met in any electrical connection.

1. There must be sufficient contact area between the two current-carrying surfaces (e.g. between wire and terminal). If this is provided, then the surface contact resistance will be minimised. There will also be a reduction in the voltage drop across the contacts and in the amount of heat generated. Note that the voltage drop is the product of the current (I) flowing through the joint or termination and the resistance (R) of the contact. The heat generated is calculated in watts and is the product of the square of the current (I^2) flowing through the joint or termination and the resistance (R) of the contact, thus I^2R watts. In practice, the volt drop and the amount of heat generated are so small that they are ignored. However, a badly soldered joint (a 'dry' joint, for instance) could cause trouble and must be rectified before damage is done, particularly to any associated insulation.
2. There must be adequate mechanical strength. This aspect is very important where there is the possibility of leads being pulled. Thus, the type of conductor termination must be considered from the point of view of mechanical damage being sustained by the joint or termination.
3. The third requirement is the ease with which a connection can be made and unmade. Electrical wires are often 'permanently' connected by soldering or crimping methods, usually where the currents to be carried are relatively low. Where, however, permanent connections are a disadvantage (e.g. in maintenance), then detachable unions are selected. These are invariably used in medium- and high-current work.

The resistance of two separable contact surfaces depends on the amount of pressure exerted to keep the surfaces together, and the conditions of the surfaces (e.g. uneven or dirty). Non-separable contacts soldered, brazed (or welded) depend on the effectiveness of the jointing method used to reduce resistance. The following are the main requirements of the IEE Regulations regarding terminations and conductor joints.

Cable terminations. All terminations of cable connectors and bare conductors must be accessible for inspection. They must be electrically and mechanically sound. No stress should be imposed on the terminals. Where two dissimilar metals are being used (e.g. copper and aluminium), care must be taken to prevent corrosion, particularly in damp situations. All insulation damaged by heat-jointing processes (e.g. soldering) must be made good. Soldering fluxes which remain acidic or corrosive at the completion of a soldering operation must not be used.

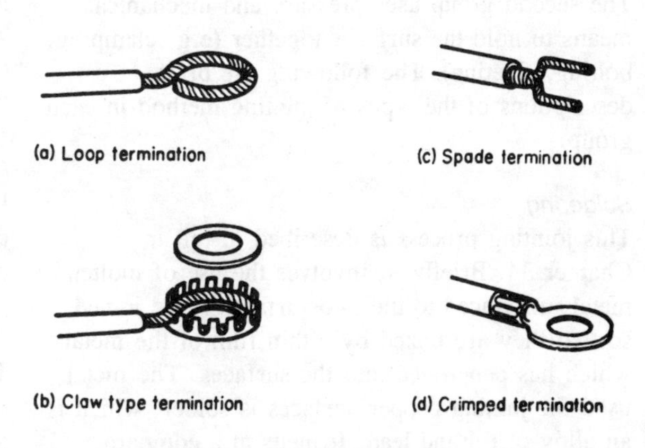

Figure 7.1. Types of conductor termination.

Joints in cables. An electrically sound joint means that the resistance of the jointed conductor should not be greater than that of an unjointed length of a similar conductor. A mechanically sound joint means that any pulling on the finished joint will not disturb the joint. A soldering joint must be mechanically sound *before* soldering. A joint which is readily accessible is one which is located usually in a box of the inspection type and the box itself must be readily accessible. The termination of a flexible cable or a flexible cord to an appliance must be done either by wiring direct onto the appliance terminals or by means of an inlet connector (e.g. kettle connector). If a joint must be made between a flexible cord and/or flexible cable, an insulated mechanical connector must be used. Non-reversible cable couplers and connectors are desirable.

Often flexible cables are required to be extended in length by the use of couplers. Their use is not regarded as being good practice, but if the situation demands it, only couplers to BS 4343 should be used. Only BS 4343 couplers are permitted on construction sites. Couplers should be non-reversible and so connected that the 'plug' is on the load side of the equipment.

Joint methods

The many methods used to join conductors may be reduced to two definite groups. The first group involves the use of heat to fuse together the surfaces of the joint (e.g. soldering and welding). The second group uses pressure and mechanical means to hold the surfaces together (e.g. clamping, bolting, riveting). The following are brief descriptions of the types of jointing method in each group.

Soldering

This jointing process is described in full in Chapter 34. Briefly, it involves the use of molten metal introduced to the two surfaces to be joined so that they are linked by a thin film of the metal which has penetrated into the surfaces. The metal used for joining copper surfaces is solder, which is an alloy of tin and lead. It melts at a comparatively low temperature. The grade of solder most suitable for electrical joints is tinman's solder (60 per cent tin, 40 per cent lead; melting point is about 200 °C). The disadvantage of soldering is that it makes the joint a non-separable contact. Soldered joints in busbars must be reinforced by bolts or clamps.

Welding

This process is sometimes used for large-section conductors such as busbars. Welding is the joining of two metal surfaces by melting adjacent portions so that there is a definite fusion between them to an appreciable depth. The heat is supplied by a gas torch or an electric arc. Again, the welded joint is a non-separable contact.

Clamping

A clamped joint is easy to make, no particular preparation being required. The effective csa of the conductor is not affected, though the extra mass of metal round the joint of termination makes a larger bulk. However, the joint or termination is cooler in operation. This method provides a separable contact. Surfaces must be clean and in definite mechanical contact. Precautions must be taken to ensure that the bolts and nuts of the clamp are locked tight.

Bolting

This method involves drilling holes in the material and has the obvious disadvantage of reducing the effective csa of the material. Contact pressure also tends to be less uniformly distributed in a bolted joint than in one held together by clamps. Spring washers are needed to allow for expansion and contraction as the material temperature varies with the current carried.

Riveting

If well made, riveted joints make a good connection. There is the disadvantage, however, that they cannot easily be undone or tightened in service.

Crimping

This is a mechanical method. For conductor joints a closely fitting sleeve is placed over the conductor and crimped by a hydraulically or pneumatically operated crimping tool. This method is very commonly used nowadays and provides a connection which is mechanically strong and virtually negligible in its electrical resistance.

Mechanical connectors
These consist of one-way or multi-way brass terminals contained in blocks made from porcelain, Bakelite, nylon, polythene or PVC. Small screws are used to make the connection. The operating temperature of the block material is important. Porcelain can be used for high operating conditions, while PVC and polythene tend to become distorted as the melting-point of 160 °C is approached. In fact, polythene is not recommended for use as connector-blocks in fixed wiring systems, accessories, luminaires and appliances. Nylon has a good resistance to deformation at high temperatures.

Termination methods

There are many methods of terminating conductors for connection to accessories and current-using apparatus. The following is a short survey of some of the more common types of termination.

Punched and notched tabs
These generally accept a solid-core small-diameter conductor. The connection is soldered.

Screwhead connection
The end of the conductor is formed into an eye using round-nosed pliers. The eye should be slightly larger than the shank of the screw, but smaller than the outside diameter of the screwhead, nut or washers. The eye should be so placed that the rotation of the screwhead or nut tends to close the joint in the eye. If the eye is put the opposite way round, the rotation of the screwhead or nut will tend to untwist the eye to make a bad, inefficient contact. Sometimes saddle washers are used to retain the shape of the eye.

Claw-type terminals
These are sometimes called segmented eyelet lugs. The conductor strands are twisted together tightly and formed into a loop to fit snugly into the circular claw. An associated brass or tinned copper washer is then placed on top. The claws are then bent over the washer.

Spade terminals
These are either preformed terminals, or made from the conductor end as follows (seven-strand conductor):

1. Strip off a suitable amount of insulation from the end of the conductor. If VRI cable, strip off the braiding and tape for a further 12 mm.
2. Take one outer strand and twist round base immediately above the insulation.
3. Separate conductor into two sets of three strands each.
4. Twist each set tightly together.
5. Form a spade end.

Lug terminals
These come in many types as shown in Figure 7.2. Connection between conductor end and the terminal's socket is made either by soldering or crimping.

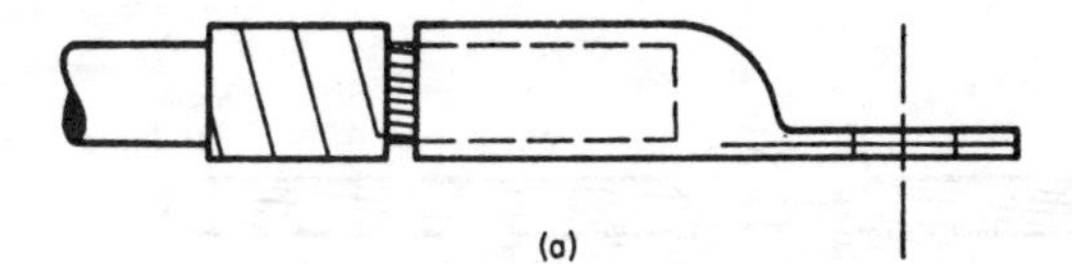

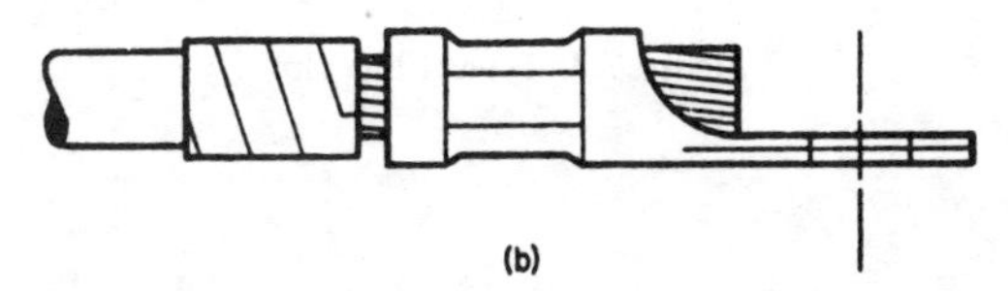

Figure 7.2. (a) Soldered lug terminal; (b) Crimped lug terminal.

Crimping. Select the correct terminal end. Strip the insulation from the cable end. Insert the wire into the open socket end of terminal and crimp using a crimping tool.

Soldering. (1) Strip the insulation back about 50 mm. (2) Tin the socket of the lug. (3) Smear both the inside of the socket and the conductor end with flux. (4) Fit the socket to the conductor. If the socket is too large, the conductor diameter should be enlarged with a tinned-copper wire binding. (5) Play the flame of a blow-torch on the socket until the heat has penetrated to the conductor. Apply solder to the lip of the socket. (6) When the termination has cooled, cut back any damaged insulation and make good. Tape can be used to protect the original insulation.

A file should never be used to smooth or clean up a soldered connection. The solder should be smoothed by wiping it with a fluxed cloth-pad while the socket is still warm.

Line-taps

These are used for making non-tensioned service or tee connections to overhead lines. They are available in a range of sizes suitable for copper conductors. A simple shroud is provided to insulate the line-tap when used on covered service cable. There are designs for use with aluminium conductors and for bimetallic connections between aluminium and copper conductors. In these instances, the shroud is filled with weatherproof sealing compound, giving protection against climatic attacks and corrosion.

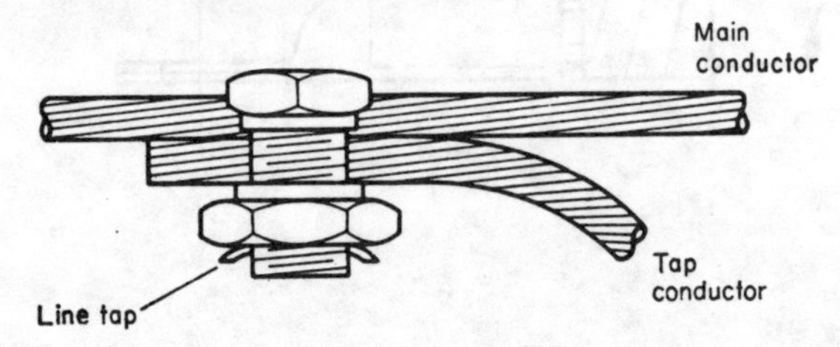

Figure 7.3. Typical line tap.

Joints and terminations on MICS cable

This type of cable consists of conductors insulated with compressed magnesium oxide and enclosed in a seamless copper sheath or tube. Generally, the ends of the cable must be sealed against the ingress of moisture by using a suitable insulated sealing compound. The complete cable termination (see Figure 7.4) comprises two sub-assemblies, each of which performs a different function: (*a*) the seal, which excludes moisture from the cable insulation, and (*b*) the gland, which connects the cable to a conduit-entry box. The seal consists of a brass pot with an insulated disc to close the mouth. Sleeves insulate the conductor tails. The gland consists of three brass components: a nut, a compression ring and a body.

There are three types of seal, each being designed for use depending on the application of the wiring system (see Chapter 8, 'Sheathed wiring systems'). Wedge-pot seals are common nowadays.

Terminating MICS cable (for use in temperatures up to 70 °C)

1. Cut the cable to the length required and allow for an appropriate length of conductor tails. The cable end should be cut off squarely. If the cable has a PVC oversheath, the PVC should be cut back before stripping the copper sheath.
2. Mark the point to which the copper sheath is to be stripped back to expose conductors.
3. Remove the sheath, using one of the recognised methods. Make sure that after

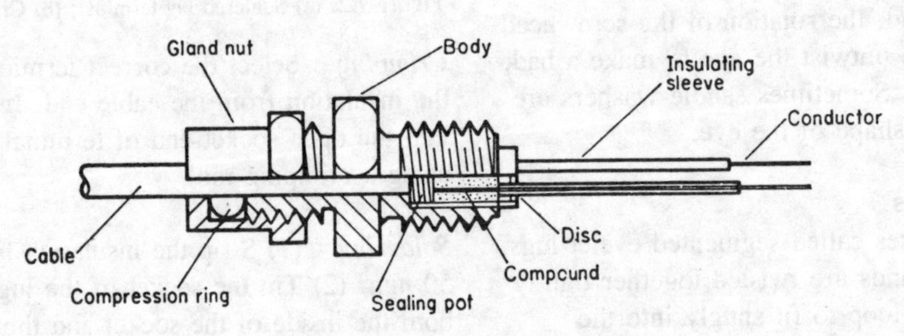

Figure 7.4. MICS cable termination.

stripping, the cable end is squared off and clean and free from burrs.

4. Clean the conductors thoroughly.
5. Slip gland nut, compression ring and gland body onto cable.
6. Screw on sealing pot (forced thread).
7. Slip disc and insulating sleeve assembly on conductors.
8. Test with insulation-resistance tester to ensure that the conductors are not touching. The reading obtained should be at least 1 megohm.
9. Press compound into sealing pot from one side.
10. Fit disc into mouth of pot. Crimp sealing pot with crimping tool and clean off surplus compound.

Conductor identification

This subject is dealt with in Chapter 6, 'Conductors and cables'.

Insulation and protection of joints

When insulating material has had to be removed to make the joint, it must be replaced with an equal thickness to ensure the final joint is electrically similar to an unjointed length of cable. This is particularly the case where insulation has been damaged, for instance where heat has travelled and softened PVC. Depending on the installation conditions in which the jointed or terminated cable is to be used, precautions should be taken to ensure that no damage can occur due to corrosion, high temperatures or the ingress of moisture.

Conductors which are to be terminated onto busbars are required to have at least 150 mm of the insulation removed and replaced with heat-resisting sleeving or tape.

Testing

All joints and terminations should be tested as follows:

1. Insulation resistance between conductors (at least 1 megohm).
2. Insulation resistance between conductors and the cable sheath if this is metal, as in MI cables (at least 1 megohm).
3. In joints and terminations intended to carry large currents, the resistance of the joint or terminal connection should be tested with a milliohmmeter. In some instances the resistance may have to be in micro-ohms to ensure that the heat generated is minimal. For example, a joint carrying 500 A with a resistance of 200 micro-ohms would produce (I^2R) 50 W of heat. It follows that higher values of resistance would produce significant amounts of heat which would then travel through the conductors and, if PVC insulation was involved, create damage.

8 Wiring systems 1: Sheathed

The wiring systems dealt with in this chapter include those whose insulated conductors are protected against mechanical damage by a sheath of either insulating material or of metal. Each system has its own advantages and limitations and each has to meet the requirements of the designer of an installation, whether it be domestic, commercial or industrial. The sheathed wiring systems which are described here are those which the practising electrician will come across in his work where new installations are being installed. However, in rewiring contracts, he/she may come across some of the older systems, such as TRS (tough-rubber sheathed) and lead-sheathed cables. These were once popular but have since been taken over by PVC-sheathed cables and MICS cables.

PVC-sheathed

PVC (polyvinylchloride) is a very common insulating material. Its main advantage is that it is cheap. It is, however, also tough and easy to work with. Because it is a thermoplastic material it tends to soften at high temperatures; the recommended working temperature (maximum) is 70 °C. Above this level, and particularly around 115 °C, the PVC starts to produce corrosive agents which will attack terminals. At the other end of the working temperature scale of PVC, the material becomes brittle and should not be installed or flexed when the air (or ambient) temperature is 0 °C.

Several grades of PVC compound are available which can be installed in a working temperature of 80 °C and as low as −30 °C. When PVC starts to burn it emits toxic vapours which are dangerous if inhaled. Some cables are available which have a low toxicity (such as 'PX') which reduces the problem.

The most common type of cable, used in domestic installations, is PVC-insulated conductors contained in a PVC sheath. The sheath is designed to protect the conductor from mechanical (or physical) damage during its installation and thereafter from any environmental hazards throughout the life of the installation. The cable can be run on the surface, though it is usually concealed behind plasterboard walls. When PVC cables are installed in attics, they should be kept away from thermal insulation material made from polystyrene as the latter has a damaging effect on the PVC. All cables are provided with an uninsulated conductor which acts as the circuit-protective conductor (CPC) which is connected to earth terminals in wiring accessories. The cables contain either two conductors (red and black) or three conductors (red, yellow and blue); these are used for two-way switched circuits.

Generally PVC cables can be installed without protection. However, there may be situations in which additional protection is required. Damage by rodents (rats and mice) is a hazard in old buildings and in farm installations, particularly where the cables may cross a 'rat run' and are gnawed away. Again, while the cables can be buried in plaster, there may be a danger from drills, screws and nails. In these situations the cables should be given additional protection by use of plastic or metal channelling or light-gauge conduit. All cable runs should be either horizontal or vertical and never diagonal, particularly under plaster.

MI copper sheath

This type of cable consists of copper conductors embedded in a densely compacted mineral insulation made from magnesium oxide. The sheath is a seamless, solid-drawn copper tube. It was originally used exclusively for installations where there was a high risk of fire and explosion, but is now used extensively for industrial applications and for fire alarm systems. Though it is expensive, it has a number of advantages which outweigh the cost. The main advantage is that it can withstand very high temperatures; continuous operating temperatures of up to 250 °C are permissible. The cable can carry much higher currents than those with other insulation materials (e.g. PVC). Its copper sheath allows it to perform the function of a CPC. And, in most situations, the sheath has a high resistance to corrosion. If, however, there is a chance that the copper could be attacked by chemicals, a PVC oversheathed cable can be used.

The main disadvantage of the MI cable is that the insulation is hygroscopic; that is, it will readily absorb moisture. This is why the cable is made off by the use of seals.

The cable is available in two types: light duty, for use up to 600 V (domestic and light duty applications), and up to 1000 V (industrial, hazardous and heavy duty applications). Normally the cable can be installed with the sheath bare. PVC oversheaths are available where there is a danger of corrosion (colour: orange), for use for fire alarm circuits (colour: red) and for emergency lighting circuits (colour: white).

Because of the cable's small diameter, the conductors tend to be closer together than in the case with other types of cable. This can present problems when the MI cable is used for circuits feeding motors and discharge lighting. Switching surges produce surge voltages which may be in excess of the supply voltage and cause an insulation breakdown. To prevent this, voltage surge protectors are used.

Mineral-insulated cables are identified by their duty rating and the cross-sectional area (csa) of the conductors: 2L1.5 means two conductors of 1.5 mm^2 in light duty cable and 4H6 means four conductors of 6 mm^2 for heavy duty applications.

MI cables are used in earthed-concentric wiring systems (TNC) where the sheath acts as a PEN (protective earth and neutral) conductors. That is, the copper sheath acts as not only the neutral or return conductor but also as the CPC. The system is used in very large installations with a view to reducing the copper cost of the job; for example, a normal two-core cable can be replaced by a single-core conductor. There are stringent requirements laid down by the Wiring Regulations before the system can be used. In particular, the copper sheath must be continuous throughout the length of run of a circuit. The sealing pots are provided with 'earth tails' which are connected together where, for instance, a conduit box is used, thus ensuring the electrical integrity of the sheath.

PVC–SWA cable

This type of cable is used for main and sub-main circuits. They consist of PVC-insulated conductors of copper or aluminium which are sheathed with stranded steel wire and then provided with an overall sheath of PVC. They are mainly found in industrial installations and are terminated by using compression glands which are terminated in such equipment as distribution boards, busbar chambers and switchgear. An earthing tag or bonding ring is provided to ensure earth continuity between the cable armouring and the metal box or panel. Gland types are available for indoor use, outdoor use and for use in hazardous areas. Shrouds are also used where there is a risk of corrosion.

Summary

1. *Flat PVC-insulated and sheathed*. For general indoor use but can be used on exterior walls with additional protection provided as necessary. The cable can also be used for overhead wiring with the cable suspended on a catenary wire by cable suspenders. If the cable is to be installed underground, conduit should be used as a duct.
2. *PVC–SWA*. For general indoor and outdoor use with precautions needed when there is a risk of corrosion of the steel wire armouring. If the cables are installed underground, protective tiles should be used directly over the cable and a cable marker placed near the top of the infill of the trench.
3. *MI cables*. For general indoor and outdoor use, with a PVC oversheath recommended when the cables are exposed to weather or corrosion. The oversheath is also needed when the cables are installed underground or in concrete ducts.

9 Wiring systems 2: Conduit, trunking, ducting

The wiring systems described in this chapter deal with basic enclosures which are designed to accommodate single insulated non-sheathed conductors. They are, of course, used for sheathed cables where additional protection from mechanical damage is required. Both conduit and trunking are available in steel and PVC. Non-ferrous conduit (e.g. copper and various alloys) is used for specialised work. Conduit is also available in a flexible form.

Steel conduit

The most common form of conduit used today is screwed steel with a welded seam or solid drawn (used in hazardous areas where there is a high risk of fire and explosion). A light-gauge conduit is also available with its use restricted to providing protection for flush PVC cable installations. Two finishes for conduit are: black enamel (dry situations) and galvanised (for outdoors and situations where dampness is present).

The main advantages of steel conduit include its ability to give conductors good protection against mechanical damage; it allows easy rewiring; fire risks are minimised; and the conduit can be used as a circuit protective conductor (cpc), though it is common practice to run a separate CPC in the conduit.

Steel conduit has a few limitations, including the problem of a build-up of condensation in situations where the temperature tends to fluctuate (this can be avoided by providing drain points in the length of run); it is expensive to erect (it is a labour-intensive system having to be measured, cut, threaded and erected before wires are drawn in); and it is liable to corrosion in adverse environmental conditions.

A full range of accessories are used with screwed steel conduit: bends, tees, draw-in boxes and adaptable boxes; all of these give a high degree of flexibility in allowing alterations to be made to an existing conduit system. In fact, once installed in the appropriate conditions a conduit system has a very long life, providing a carcass which can be completely rewired over several periods in the life of a building.

Conduit is available in four sizes (measured on its outside diameter): 16, 20, 25 and 32 mm; the normal length is 3.75 m.

The Wiring Regulations restrict the number of conductors that can be drawn into conduit. This is to allow for ventilation of current-carrying cables, to allow for removal and replacement of conductors and, in some cases where the existing conduit capacity is not up to its limit, to allow new circuits to be drawn in. Although manufacturers' conduit bends can be obtained, on site conduit is bent on a bending machine which ensures that the minimum internal radius of the bend is equal to $2\frac{1}{2}$ times the outside diameter. If, however, a bending block (made from a baulk of wood) is used it is essential that the internal radius of the bend complies with the dimension given.

PVC conduit

Where appropriate, PVC conduit is a popular, and inexpensive, alternative to steel conduit. It is available in both light and heavy grades and does not need to be threaded unless so specified by the job. The conduit is available as rigid, semi-rigid, flexible round (for surface and embedded work) and in an oval shape (for switch drops). Grades of PVC conduit include super high impact, standard impact, and high temperature (up to 85 °C).

Because the expansion rate for PVC conduit is around five times that of its steel equivalent, expansion couplers are needed in long runs (at every 8 m). Where the conduit is to be used in damp situations, a special non-setting adhesive ensures a seal which allows for movement as temperatures fluctuate.

Note: The adhesives used with PVC conduit will give off fumes which create a health hazard and thus the electrician is advised to ensure that the working place is well ventilated. Contact with the skin should also be avoided.

A wide range of conduit accessories is available, similar to those for steel conduit.

Flexible conduit

Flexible metallic conduit is often used to make a suitable connection between a rigid conduit system and, for example, a motor which may be required to be moved for belt tensioning, belt removal and replacement. Several types are available. A separate CPC is needed, run either inside the conduit or externally.

Trunking

Trunking is a fabricated casing for conductors and cables, generally rectangular in shape with a removable lid which allows the conductors to be laid in rather than be drawn in as is the case with conduit. It is used where a large number of conductors are to be carried, or follow the same route. Both steel and PVC trunking are available, with a wide range of such accessories as bends, tees, flanged adaptors, risers and reducers.

The variety of trunking includes plain section, compartmented, skirting, bench, floor trunking, and busbar trunking. Trunking is not necessarily a complete wiring system in itself and is thus associated with conduit and MI cables to allow connection to wiring accessories and their mounting boxes.

Finishes on steel trunking include grey enamel, galvanised and silver enamel on zinc-coated mild steel.

Compartmented trunking allows wiring at different voltages to be segregated but carried within the same unit run. This prevents services at one voltage accidentally becoming live to a higher voltage in the event of a fault.

Skirting trunking is used in offices where the services (socket-outlets, etc.) can be sited on the perimeters of rooms.

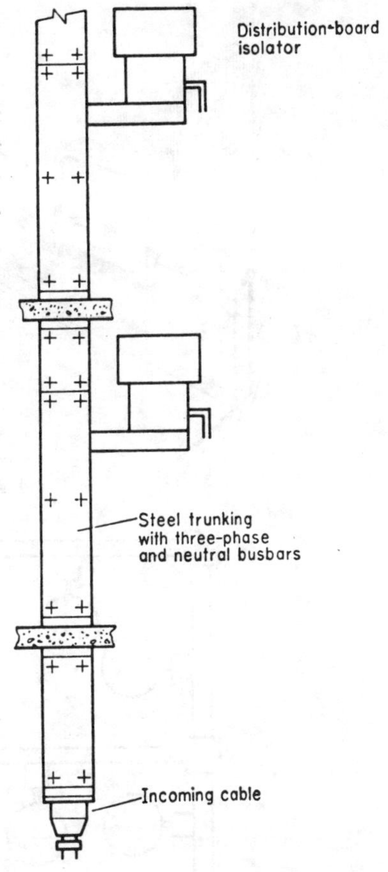

Figure 9.1. Distribution by rising mains.

Bench trunking is commonly found in schools and laboratories where access to a large number of socket-outlets is required. As the name implies, the trunking units are mounted on benches.

Floor trunking is an alternative to skirting trunking. There are three types: underfloor (where the trunking is set in a concrete floor with access only at junction boxes), flushfloor trunking (with the lid mounted flush with the floor surface) and flushduct trunking (where the lid is mounted flush with the screen) and a finish (such as parquet wood or tiles) is placed directly onto it.

Busbar trunking is basically plain-section trunking containing fixed copper or aluminium bars. Access to the busbars is made by means of tap-off boxes. It is often used in workshops where machinery or equipment may be shifted to different positions in the same area. Down drops

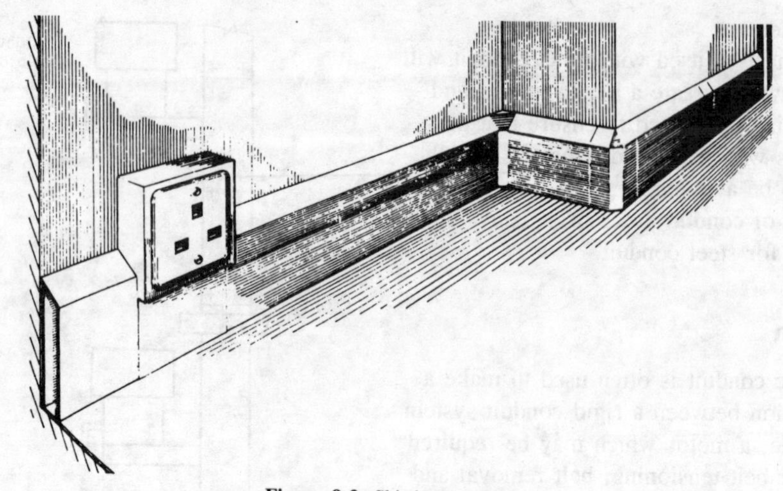

Figure 9.2. Skirting trunking.

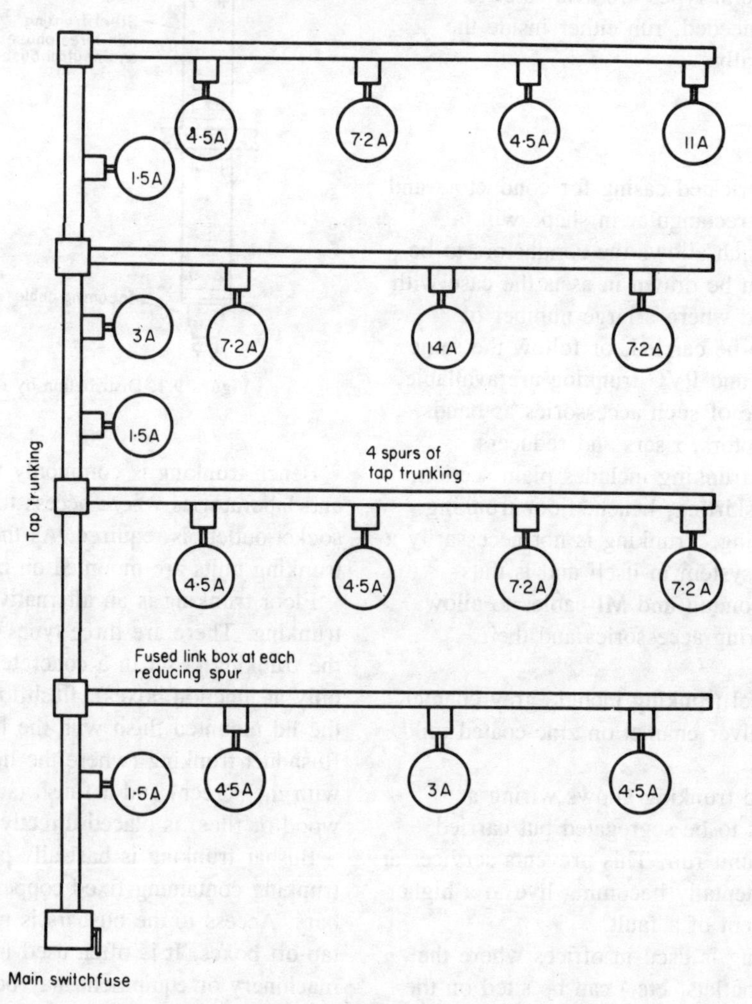

Figure 9.3. Distribution for motors using trunking.

are then available from the overhead busbar trunking tap-off boxes, via rigid or flexible conduits.

A variation of busbar trunking is the cable-tap trunking which is plain-section trunking containing cleats which support large csa PVC-insulated cables which form a useful ring-main supply from which fused tap-off boxes provide supplies for smaller circuits.

PVC trunking is made from high-impact PVC and is an inexpensive alternative to the steel equivalent. As with PVC conduit, allowances must be made for expansion of the trunking in high ambient temperatures.

In both steel and PVC trunking, the Wiring Regulations require a limit to the number of cables which can be accommodated to allow for adequate ventilation of heat generated by current-carrying cables and to allow for easy withdrawal of circuit conductors. A separate CPC must be run for each circuit run in PVC trunking.

Mini trunking is a small-section high-impact PVC with a clip-on lid. Used for surface work it is unobtrusive and is adapted to enter into ceiling roses, socket-outlets and switch boxes. It is popular for use in temporary installations such as Portacabins on building sites but can also be used effectively in domestic installations where the decoration must not be disturbed.

Ducting

Ducts are simply passages provided by builders in the structure of a building to allow cables to run from points of supply to their terminations. Ducts can be rectangular channels covered by steel lids, trenches in concrete with covers or simply pipes. All cables run in ducts must be sheathed or armoured.

Cable tray

Although not quite a wiring system, cable tray is used to carry heavy armoured cables and multi-runs of MI cables. It is widely used in industrial installations where normal cable runs may be obstructed by pipework and other structural features. The tray is basically a flat metal sheet with perforations and either a simple turned flange or a return flange for greater strength. A range of finishes are provided to meet installation conditions: galvanised, primed with red oxide or yellow chromate, plastic-coated and coated with epoxy resins (resistant to acids and virtually non-flammable). For very heavy cable runs, cable ladders are used.

10 Wiring accessories

There is an extremely wide range of wiring accessories now available, most of which the practising electrician will install at some time or another. These include switches for lighting, water-heaters, socket-outlets, cooker units, dimmer switches, ceiling roses and cord outlets.

Switches

The most familiar switch is that used to control lighting circuits. Most are rated at 5/6 A, but ratings at 15 A are also available. They are 'single pole' which implies that they must be connected in the phase conductor only. Care should be taken that lighting switches are designated for use on inductive circuits, particularly when they are used to control fluorescent lighting. This is because such circuits take 80 per cent more current than the lamps' wattage might suggest. If switches are not rated for inductive circuits, they must be derated by 50 per cent.

Three types of switches are available: one-way, two-way and intermediate, each for the control of a particular circuit arrangement. Often a number of switches are contained within the same switch unit: two-gang, six-gang, etc. This allows the control of a number of different circuits from one position. One special type of switch is the 'architrave', which is mounted on door architraves.

Ceiling switches are rated at 6 A, 16 A and 40 A and are used for either lighting or wall/ceiling mounted heating appliances in bathrooms and are of the pull-cord type.

Switches for water-heaters are of the double-pole type and rated to carry 20 A. Other ratings for double-pole switches are 32 A and 45 A, the latter being used to control cooker circuits where no socket-outlet is required in cooker-control unit.

Dimmer switches are used to allow control of the level of lighting from a luminaire. Watertight switches are designed for outdoor use while splashproof switches are found in situations where water is present, such as in shower rooms.

Most switches tend to be made from moulded plastic, but metal-clad versions are available for industrial use. Some switches for domestic installations can be finished in satin chrome or polished brass.

Lampholders

Cordgrip lampholders are used for pendant luminaires and are fitted with 'skirts' to provide extra safety when a lamp has to be changed. This is a requirement when pendant luminaires are used in bathrooms. For filament lamps rated up to 150 W, the connection is bayonet cap (BC). Higher ratings require an Edison screw (ES) lampholder in which the centre contact must always be connected to the phase conductor.

Battenholders are used for wall or ceiling mounting and can be 'straight' or 'angled'.

Ceiling roses

Ceiling roses are used with cordgrip lampholders for pendant luminaires. They must not be used in circuits exceeding 250 V and must not have more than one outgoing flexible cord unless designed for multiple pendants. They are often provided with a loop-in terminal ('live') which is required to be insulated against the possibility of receiving a shock from direct contact.

Cooker outlet units

These are units designed to accommodate the cooker supply cable from the control unit. The cable is terminated at terminals from which the cable going to the cooker is connected.

Shaver supply units

These are now commonly fitted in bathrooms and provide both 115 V or 240 V. They are often incorporated into a lighting unit for fitting above mirrors.

Cooker-control units

These can be of the double-pole type incorporating a 13 A socket-outlet, or simply a double-pole switch rated at 45/50 A. In a kitchen provided with an adequate number of socket-outlets there should be no need for the socket-outlet in the control unit as there is a danger of flexible cords (e.g. supplying a kettle) trailing over hot cooker hobs.

Socket-outlets

These take 13 A fused plugs and can be non-switched, switched or switched with a pilot lamp. Socket-outlets intended to supply electrical equipment outside the house must have residual-current device (RCD) incorporated so that should an earth fault occur, the supply will be cut off (RCDs operate at 30 mA but can trip at as low as 10 mA). Associated with socket-outlets connected to ring-main or radial circuits are fused or fused/switched connection units. The former are used where the appliance has its own switch. (*Note*: A heating appliance may have thermostatic control, but the thermostat does not act as a switch.) Switched/fused units are used where the connected appliance has no switch control. It should be noted that 2 A, 5 A and 15 A round-pin socket-outlets are available for special purposes. Connection units can also be obtained with a cord outlet.

Outlet plates

These accessories include: outlet plates with 3 A fuse used for mains-operated electric clocks; outlet plates for telephones; and TV coaxial socket-outlets. While the latter two are not directly associated with the electrical installation, contractors are often required to install these services while the building is being hard-wired.

Mounting boxes

These are designed to contain wall-mounted accessories such as switches and socket-outlets and are either moulded plastic or metal. The entries to the boxes are by means of 'knock-outs' which can accommodate conduit, flat cable or MI cable. Depending on the accessory, the box depths range from 16 to 32 mm. The boxes are also designed for surface-mounted accessories or flush-mounted accessories. All boxes are normally provided with an earth terminal for the CPC.

Grid-switch system

This system allows the control of a large number of different circuits from one position. It is often found in commercial premises, restaurants, public bars, offices and industrial situations. The system consists of a mounting box, an internally fixed grid which then accepts the switch or switches and a cover plate. The system ranges from one-gang units to twenty-four-gang units. The accessories include switches, bell pushes, indicator lights and key-operated switches.

Industrial socket-outlets and plugs

There are two types available: BS 196 and BS 4343 with current ratings from 16 A to 125 A. Colour identification is used for different voltages: yellow — 110 V; blue — 240 V; red — 415 V. The socket-outlets are designed so that the earth contact position with respect to a keyway is varied for each voltage rating to ensure that equipment of a given voltage cannot be plugged into the wrong supply.

Installation hints

The following list is based on the requirements of the Wiring Regulations relating to accessories.

General. All mounting boxes must be securely fixed, with no sharp edges on cable entries, screw heads, etc. which might cause damage to cables and wires. Cable sheaths should be fully entered into the box. All conductors are required to be correctly identified with bare CPCs sleeved with green/yellow sleeving. All terminals should contain all the strands of conductors and be tight.

Lighting switches. Single-pole switches are to be connected in the phase conductor only which must also be correctly identified by colour. All exposed metalwork (e.g. the metal switch plate) is required to be earthed. In a bathroom, the lighting switch must be of the pull-cord operated type, or else mounted outside the bathroom door. If a switch is not rated to carry the current of inductive circuits (e.g. fluorescent luminaires) it must be derated by 50 per cent.

Ceiling roses. They should not be connected to a voltage more than 250 V and should not have more than one flexible cord coming from it. Loop-in terminals must be insulated and the rose should be suitable to take the weight of a luminaire.

Socket-outlets. They should be mounted at a height above the floor or working surface which is convenient to the client. Note that in premises where disabled people live and work, the socket-outlets should be located at a level which allows them access without strain. Correct polarity must be observed in wiring socket-outlets, with an earthing tail provided between the s/o earth terminal and the terminal provided if a metal mounting box is used. Socket-outlets are not allowed in rooms containing a bath or shower. They should be more than 2.5 m away from a shower cubicle in a room other than a bathroom.

Cooker-control unit. These must be located within 2 m of the cooking appliance.

11 Installation methods

General considerations

All electrical installations are required to be properly designed with a number of important factors taken into consideration. These factors have a bearing on the type of wiring system to be installed, its flexibility, its maintenance aspects, the working environment and the degree of safety provided for the users of the installation. The Wiring Regulations, Statutory Regulations and other recommendations all have an influence on the designer's approach to satisfying a client's requirements, whether the latter be the owner of a domestic dwelling, a small shop, a supermarket, a school, an office block or a factory. Each has its own specific aspects which both the designer and installer (the contractor's electricians) have to provide to ensure that the final installation meets the highest standards. Thus the golden rule is: 'Good workmanship and the use of proper materials will meet these standards'.

The type of building. This can range from a domestic dwelling to a large factory, from a suite of offices to farming premises, from a small shop to a shopping mall. All these incorporate a wide variety of building materials with which the electrician must be familiar. The users of the installation also have to be considered.

On the face of it, a domestic installation may seem straightforward. But if the installation is to cater for a disabled person, then it must be designed in such a way that the person is not inconvenienced. Farm installations must take into account the fact that even 25 V is lethal to animals. A shopping mall installation may have to be designed to cope with vandalism. Some buildings may be of a temporary nature. Others may at some time in the future be extended. While the installation designer may have done his/her job correctly, it is the installer or electrician who comes face to face with the practical aspects which at times may require to be notified to the client's agent if something might not seem to be 'quite right'.

Installation flexibility. A flexible installation is one which can be altered or extended easily without undue disturbance of the fabric of the building. Some buildings may have different tenants during their lifetimes, all of whom may have a wish to alter the installation to suit their needs. Even domestic installations may require to be extended as a family grows larger, or as rooms are altered to other purposes. Thus initial planning, if done thoughtfully, will save a client extra cost.

Working environment. Electrical installations can find themselves coping with dust, fumes, chemical vapours, dampness, weathering, and high or abnormal temperatures. All these operating conditions can have a deleterious effect on both wiring systems and electrical equipment. It is thus essential that consideration be given at the initial planning stages of an installation of the conditions expected in the installation. It is again the collaboration of the designer and the installer which will ensure a problem-free installation.

Maintainability. This is important where the installation is intended to have a long life without disturbance. The design requirement here is that the installation can be easily tested periodically and that repairs can be readily and safely carried out. It is also essential that any safety and protective measures remain effective during the life of the installation.

Safety. All electrical installations are required to have an in-built safety factor, to protect the users against the risk of electric shock, fire and burns.

While this can be done at the design stage, much depends on the installer's approach to the work, ensuring that wiring is correctly installed and that equipment is correctly connected. An indication of the number of common faults which appear in installations can be seen in *Snags and Solutions* published by the National Inspection Council for Electrical Installation Contracting. These faults have been detected over many years by the Council's inspecting engineers — and they still occur!

Installation practice

Conduit (metallic)

The following is concerned with heavy-gauge screwed conduit. Light-gauge is available but is limited in its use in installation work to the protection of sheathed cables from mechanical damage. In any case, they must not be used as a circuit protective conductor (CPC). The heavy-gauge type of conduit is available as seam-welded or solid drawn. The former is the more commonly used while the solid drawn is, apart from being expensive, mainly used for special installations such as in gasproof or flameproof situations.

The conduit wiring system is a two-stage system: the conduit must be fully erected and supported before any cables are drawn in. It is thus a 'labour-intensive' system which contributes a significant element to the cost of a conduit installation. The advantages, however, far outweigh the cost because of its long life (in the appropriate environmental conditions), its ability to allow rewiring, replacement of cables and adaptability to other systems such as trunking. That the conduit can be used as a CPC is an advantage, though it is current practice to run a separate CPC within the conduit. The two essential requirements of a conduit system are: it must be properly constructed and it must be electrically continuous.

There are a number of practical points to be observed. All conduit must be cut squarely and all burrs removed. Threading should be done carefully and no damage should occur to the finish, particularly if the black enamel type is used. No exposed threads are allowed except where running couplers are used; in this case the threads should be treated to prevent rust and corrosion. The radius of site-made bends should be not less than $2\frac{1}{2}$ times the outside diameter of the conduit. All entries into enclosures should be bushed to prevent abrasion of insulated conductors. All unused entries should be blanked off using brass plugs.

All covers of boxes should be in place and securely fastened. If these boxes are located under floors, trap-doors should be provided for access at a later date. All bushes, couplers and accessories must be securely tightened. In long horizontal runs of conduit, where there is a risk of condensation collecting inside the conduit, drainage points should be provided. The capacity of the conduit must not be exceeded. This is to allow space within the conduit for ventilation of heat from the conductors and to allow easy withdrawal and replacement of conductors.

The number of conductors allowed in conduit depends on whether the run is short and straight (maximum 3 m) or whether the run is longer and incorporates bends or offsets. Tables in the 'Guidance Notes' give details of how to calculate the maximum number of cables for any size of conduit.

Each size of cable (whether solid or stranded conductors) is allocated a 'cable factor'. These are then added together and the total cable factor compared with a 'conduit factor'. The cable factor must be equal to or less than the conduit factor.

Example. A conduit run of 7 m incorporating two bends is to accommodate 4×1.5 mm^2, 3×2.5 mm^2 and 2×4 mm^2 cables.

Cable factor for 1.5 (solid) $= 27$
Total cable factor $= 4 \times 27 = 108$

Cable factor for 2.5 (solid) $= 39$
Total cable factor $= 3 \times 39 = 117$

Cable factor for 4.0 (stranded) $= 58$
Total cable factor $= 2 \times 58 = 116$

Therefore the sum total $= 341$.

Reference to the conduit factor tables will indicate that these cables can be accommodated in 32 mm conduit (conduit factor $= 404$).

In general the longer the run, or the more bends in the run, will tend to increase the conduit size. For instance in the example, if only one bend was incorporated, the conduit size becomes 25 mm.

Drawing in cables is carried out by using a draw-in tape made from steel or nylon which is fed into the conduit. A draw wire is then attached to the tape and itself drawn in. The cables are then firmly joined to the draw wire and fed into the conduit in parallel to prevent them crossing each other. This is a two-man operation with one person pulling and the other feeding. It is advisable to use a cable drum stand so that the cable from each drum can be pulled directly off its drum and not allowed to spiral off, causing twists in the cable.

The methods of fixing conduits included the clip (a half-saddle), saddle, spacer-bar saddles and distance saddles, the latter being used to take the conduit off the wall by about 10 mm. Other types of saddles are available for multiple runs and girder clips where conduits are run across girders. The 'Guidance Notes' stipulate distances between conduit fixings to ensure the conduit is securely erected.

Inspection fittings (tees, elbows, bends) of the channel type are not generally recommended except in situations close to the conduit-entry points. They do not have sufficient space for drawing in cables. Draw-in boxes not only look better but have ample capacity, not only for drawing in cables but to accommodate a turn or two of cable to provide some 'slack' in the conduit run.

Where conduits have to be concealed, they are installed while the building is being erected. They can be buried in floors and walls in such a manner that the cables can be drawn in after the building work has been completed. Conduits running under floorboards and across joists should be accommodated in channels cut in the joists sufficient only to accommodate the conduit.

The use of solid elbows or tees is restricted to positions immediately adjacent to a conduit terminal box, a luminaire or conduit inspection fitting and in any case must be located not more than 500 mm from a fitting which affords permanent access.

Where conduit runs pass through walls or ceilings, the surrounding holes must be made good with fire-resisting material, such as cement, as a fire precaution.

Galvanised conduit can be used as part of a system carrying cables from one building to another, provided that the maximum length does not exceed 3 m and the height is 3 m minimum, but higher if there is traffic. The conduit must be one unjointed length and be securely fixed at both ends of the span.

Conduit (PVC)

Many of the requirements for metallic conduit also apply to PVC conduit. Some requirements are, however, peculiar to PVC. The provision of expansion couplers is essential in long runs where abnormal temperatures are encountered. Separate CPCs are required. They must not be installed where they are likely to come into contact with materials which might cause them damage, such as creosote and oil, unless the PVC is designed to withstand these contacts. If a plastic box is used to support a luminaire the box must be suitable for the suspended load at the expected temperature.

Compared to metallic conduit, PVC conduit is easy to work with and virtually the same range of system accessories and supports used for metallic conduit is available for PVC. Bending can be done on site using bending springs, though it is essential that care is taken when working with PVC in cold weather. Expansion couplers are fitted every 6 m of the run. Jointing PVC conduits is by means of a push fit and sealed with PVC solvent adhesive used sparingly.

Flexible conduits

The purpose of this conduit is to provide a flexible connection between a fixed conduit or trunking installation and some types of electrical equipment (e.g. motors) where there is a need for the equipment to be moved within small limits of its mounting position. The conduits are also used to absorb vibration to prevent it being transmitted to the rigid installation. Types include metallic, plastic flexible and reinforced, which is a heavy-duty double-walled conduit with a spiral wire reinforcement. Adaptors are used at both ends of the conduit length. A separate CPC is required by the Wiring Regulations. Flexible metallic conduit should be of the waterproof type.

Conduit summary

All metallic and PVC conduits must be erected and securely fixed before cables are drawn in. Any inspection (channel type) fittings must always remain accessible as must all draw-in boxes (by using traps whose positions must be marked). Solid elbows and tees can be used only as permitted by the Wiring Regulations. All unused entries must be blanked off or plugged. Metallic conduits which are run with other service pipes (e.g. in factories) should be painted light orange, the recognised colour to indicate an electrical service. In situations where condensation might occur, drainage points must be provided. The minimum internal radius of site-made bends must not be less than $2\frac{1}{2}$ times the outside diameter of the conduit. The permitted number of cables (e.g. the total of the individual cable factors) must not be exceeded (that is, must not exceed the conduit factor).

All conduit runs must be segregated from other metalwork; if this is not possible the conduits must be bonded to the extraneous metalwork. If the conduit is being used as a CPC, all the joints must provide satisfactory earth continuity. The resistance of metallic conduit is 5 milliohms/m, which includes both unjointed lengths of conduit and lengths incorporating boxes, couplers, etc. In damp situations, the conduit should be galvanised or sherardised. Phase and neutral conductors must be bunched in the same conduit (if metal).

In the case of PVC conduit, the upper and lower temperature limits should be observed. There must also be provision for expansion and contraction in abnormal working temperatures. Boxes intended to suspend luminaires must be suitable for the expected temperature.

Flexible metallic conduit must have a separate CPC run inside or outside and it must always be adequately supported.

MI cables

Perhaps the most important aspect of working with MI cables is the need to ensure that dampness has not entered into the cable insulation. Two other aspects are the danger of overworking the copper sheath in bending and making offsets, and the protection of the copper sheath with a PVC oversheath in situations where there is a danger of corrosion. A point also to bear in mind is that there are two grades of cable: light duty, up to 600 V, and heavy duty, up to 1000 V. The former is used for domestic and light commercial installations, the latter mainly for industrial installations.

The small overall diameter of the cable allows it to be installed in walls without the need for deep chases, though the cable may need a PVC oversheath to avoid any possible interaction between the copper sheath and the plaster or cement (the latter can contain corrosive salts). The minimum internal bending radius must not be less than six times the overall diameter of the cable. The spacing of clips and saddles should conform to the requirements of the Wiring Regulations.

Before the seal (screw-on pot or wedge-type pot) is finally crimped the insulation resistance between the cable cores and the sheath must be measured and should be a minimum of 1 megohm. The cores or conductors must also be identified by colour (usually tape) after the conductors have been checked by means of a continuity test (using the ohms scale on the instrument).

In industrial installations, MI cables are often carried on cable tray in multiple runs, for which multi-way saddles are available. Such saddles can also be site-made using narrow copper strip which is available with PVC covering as are clips, saddles and spacer-bar saddles.

Because the MI cable system is used with conduit fittings, pots are available with earth tails which, joined together, provide a good electrical (copper) continuity between sections of the cable. This is particularly important if the MI cable is being used in earth-concentric wiring systems (TNC), where the sheath acts as a PEN conductor (that is, it is the neutral or return conductor and also the CPC).

MI cable summary

MI cables may require surge voltage protection when used to feed certain types of inductive loads.

These voltages, which are generated when the inductive loads are switched off, can be far in excess of the normal supply voltage and will cause the breakdown of the cable. Inductive loads include motor circuits and discharge lighting circuits (e.g. fluorescent lamps).

Earth-tailed pots must always be used if there is any doubt about the effectiveness of contact between the cable gland and an enclosure. If the enclosure is made from insulating material, the use of earth-tailed pots ensures a facility for making a 'through-joint' in the protective conductor.

When the MI cable is being used for Category 3 circuits, the cable should have a PVC oversheath coloured to indicate the function of the circuit: red for fire alarms and white for emergency lighting.

The correct polarity at socket-outlet terminals must be verified. This is particularly the case where black core sleeving is used. Recently, red coloured sleeving has come on the market which can be used and does away with the need for coloured tape for the phase conductor.

Note that there are two installation conditions recognised by the Wiring Regulations: sheath bare and exposed to touch, or sheath bare and not exposed to touch. There is, further, the situation in which the cables can be clipped direct to a surface or in 'free air'. In addition it is essential to recognise the working sheath operating temperature: 70 and 105 °C, which governs the material used for filling the sealing pots.

Trunking

Trunking, whether metallic or plastic, is basically a large-volume channel for housing cables. The range of trunking types available today makes it a versatile wiring system and in its various forms can be found in domestic, commercial and industrial situations. In its metallic form it is commonly used in industrial premises where it is associated with metallic conduit and MI cable systems.

Trunking can be fixed directly to a wall (using round-head screws and washers to prevent damage to cables) or can be run overhead supported at intervals by the bottom members of the roof trusses.

The spacing factor for trunking must always be observed. As in the case with conduit, cables are given a factor. The total sum of the cable factors must not exceed the trunking factors, which are quite generous in view of the large capacity of trunking.

The wide range of trunking accessories available cut out the need for much work on site in making purpose-made accessories, though it is useful for trainee electricians to gain some experience in working with trunking.

Since trunking offers mechanical protection for cables, single insulated conductors can be installed. This is done by simply laying the cables for each circuit in the trunking either freely or by supporting the circuit conductors with cable-retaining straps. Care should be taken to ensure that the cables are not twisted, so that they can be lifted out easily should they need to be replaced. Vertical runs of cable in trunking require to be adequately supported by using pin racks which support the cables by friction. They should be fitted at 1.5 m intervals.

When trunking is mounted vertically, the Wiring Regulations require that fire barriers be fitted at intervals not exceeding 5 m. This is to ensure that the trunking does not allow fire to spread, and also helps to reduce the rise of temperature at the top of the trunking run.

Metallic trunking can be used as a CPC, though separate CPCs are recommended to be run with the circuit conductors. The sections of trunking are connected together (electrically) by means of earth

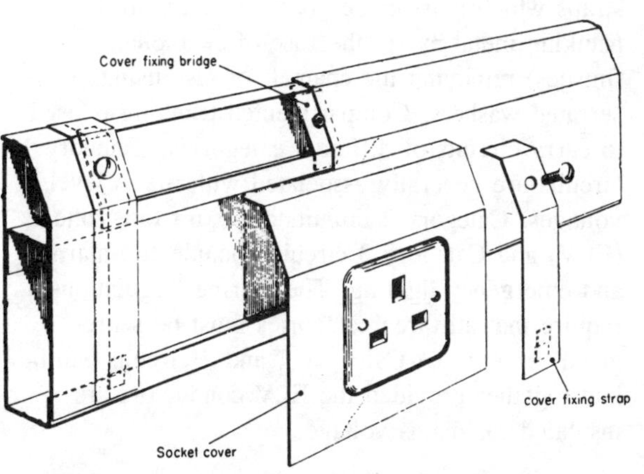

Figure 11.1. Detail of skirting trunking and accessory.

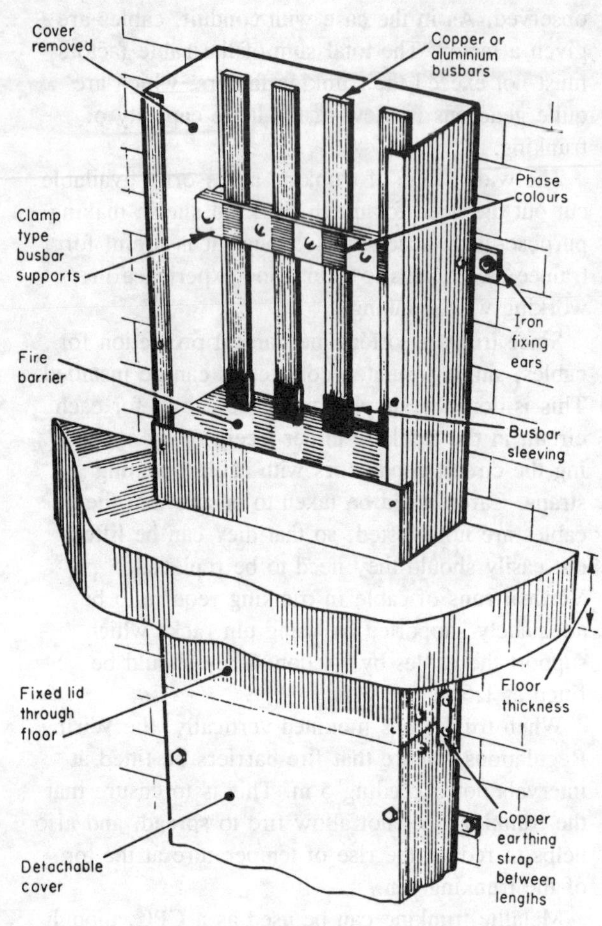

Figure 11.2. Detail of rising mains.

straps which must make good contact with the trunking metal by (in the case of enamelled finishes) removing the enamel, or else using serrated washers. Compartmented trunking is used to carry circuits of different categories. Category 1 circuits are generally associated with mains-level voltages. Category 2 circuits are extra low voltage (ELV) and Category 3 circuits include fire alarms and emergency lighting. The Wiring Regulations require that all three categories must be segregated, except that Category 1 and 2 circuits can be laid together provided the ELV conductors are insulated for mains voltage.

Trunking summary

If a socket-outlet is fixed in metal trunking, its earthing terminal must be connected by a protective conductor to a similar terminal in the trunking. If a Category 3 circuit is to be installed in trunking with other categories, the cable used must be MI which then has an 'exposed to touch' rating. Any type of cable run in vertical trunking must be supported at 5 m intervals. The capacity of trunking must not exceed 45 per cent; this is calculated by comparing the sum of the cable factors against the trunking factor.

All trunking must be securely fixed and be of a finish suitable for the installation conditions. Barriers are required to prevent an excessive temperature rise in vertical runs and barriers are also needed to prevent the spread of fire. Any holes made in walls or ceilings (where the trunking runs from one area to another) must be made good by using non-combustible material.

Where the trunking is to be used as the protective conductor, all joints must be mechanically sound and electrically continuous. The resistance of trunking is 5 milliohms/m; this figure applies whether the trunking is continuous or made up from sections and joints.

Cable tray

Cable tray is mostly found in large commercial and industrial installations. It is basically a steel tray with slots provided for fixing cable supports. The standard type has a simple flange. Heavy duty types have a return flange to provide extra strength. Finishes include galvanised, paint primed (red oxide or yellow chromate), and plastic coating. A range of accessories enables the route of the tray to be changed. All cable trays must be adequately supported. Sections are joined together by patent couplers and fasteners, 'socket' joints or by using fish-plates. Though not used as a protective conductor, all sections in the run of a cable tray route must be bonded together. Cables are fixed to the tray by means of saddles, spacer-bar saddles or, in the case of heavy cables (e.g. PVC wire-armoured), cleats.

12 Current-using apparatus

Electricity has been one of the most important factors in the social progress made by this country over the past half-century. It affects health, education, housing, standards of living and industrial and agricultural progress. Its universal application is seen in the many different types of current-using appliances found in the home, in the office and in the factory making it possible to perform tasks easily, safely and efficiently. This chapter deals with some of the more common current-using equipment, their applications and their general installation requirements.

Water-heating

Electric water-heating did not get under way until the early 1930s. When it did begin to make some impact there was a great outcry from the solid fuel and gas interests about inefficiency in the use of electricity for heating water. However, it was shown that raising the temperature of water by means of electricity is an almost 100 per cent efficient method.

The main feature of the electric element used for water-heating is that it is a resistive conductor, insulated and protected from direct contact with the water. The element material is usually of nickel-chromium alloy and wound in the form of a spiral. The insulating material is compressed mineral-oxide powder which can withstand the high temperatures produced by the element. The sheath is of copper or stainless steel. Elements made in this form are able to withstand rough usage. Other types of element consist of spirals of nickel-chromium alloy enclosed in porcelain insulators and mounted on a central rod. Their construction is such that they can be withdrawn for inspection and repair.

Electric kettle. The electric kettle is probably the most common of all electric appliances. Made in a large variety of patterns, the body is either metal or plastic. The heating element is spiral-wound resistance wire, insulated with magnesium oxide and sheathed. The element, being preformed and self-contained, can be fitted and removed easily from the body by means of a screw thread. A gasket is placed between the element and the body to make a water-tight joint. Loadings vary from 60 to 2400 W. Connection is made by means of an appliance connector provided with a sliding earth-contact. Protection against damage, due to accidental operation while dry, is provided by a safety device containing a bimetallic strip and small heating element. Sometimes an auto-ejector device is incorporated which pushes the appliance connector out of the kettle to disconnect the element from the supply should the kettle boil dry.

Free-outlet water-heater. This type is sometimes known as a non-pressure type. Generally, it has a cylindrical container with inlet and outlet pipes.

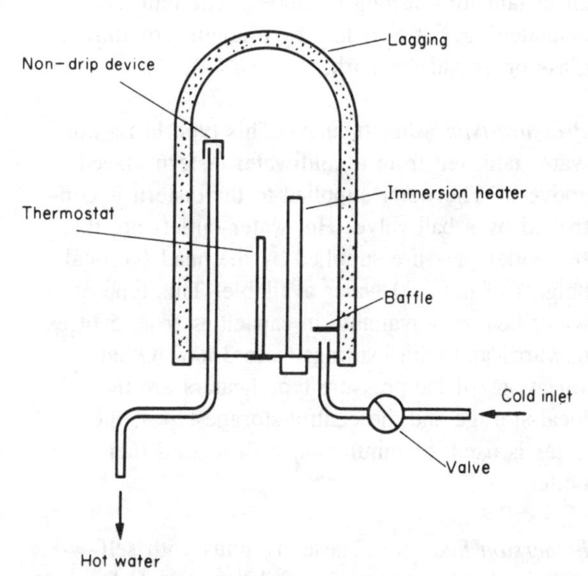

Figure 12.1. Non-pressure-type water heater.

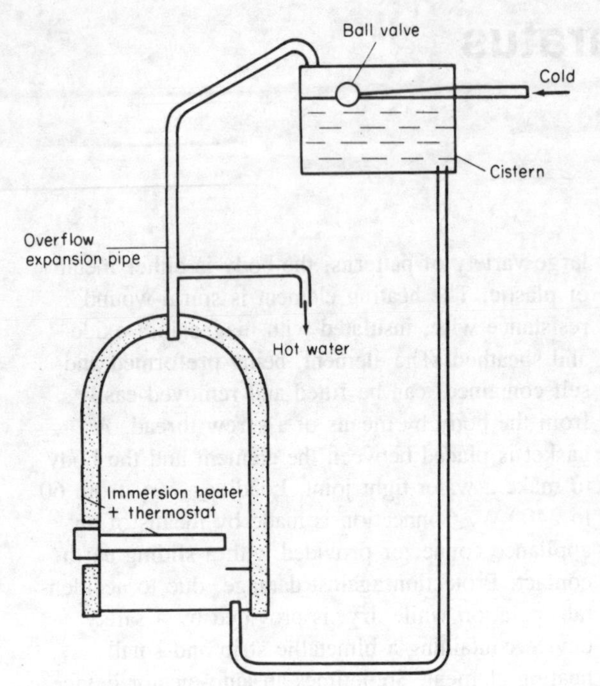

Figure 12.2. Pressure-type water heater.

The inlet is connected to the water mains through a valve; the outlet is left open. A heating element and a thermostat are located in the bottom of the container; the latter controls the temperature of the water in the tank. This type of heater is used to give small quantities (6–12 litres) of hot water in an instant for washing-up duties. The tank is insulated against heat loss by a lagging of fibreglass or granulated cork.

Pressure-type water heater. This type has a hot-water tank fed from a cold-water cistern placed above it. The water supplied to the cistern is controlled by a ball-valve. Hot-water outlets are thus fed under pressure supplied by the head (vertical height) of the cold water available. This type of water-heater is available in capacities from 5 litres upwards and with loadings from 3 kW. Other variations of the pressure-type heaters are the local-storage and the central-storage types; the latter is used in conjunction with a solid-fuel boiler.

Immersion heaters. These are units with self-contained heating elements and thermostats for use

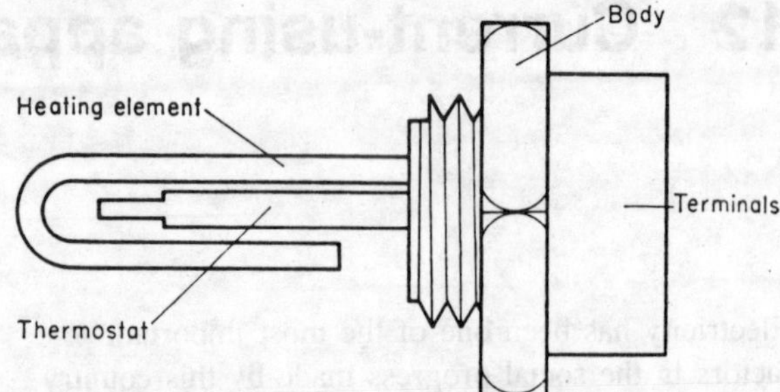

Figure 12.3. Typical water-heater, immersion type.

where the main heating is provided by a solid-fuel boiler and only supplementary heating is required by electric means. The heater consists of a sheathed-wire element, enclosed in a copper sheath from which it is insulated by compressed magnesium oxide powder. The tank must, of course, be efficiently lagged to cut heat losses.

The immersion elements are normally fitted into the top of the tank which allows the water to be heated in bulk. Elements which are located horizontally are usually associated with dual-element tanks, where the top element will heat a small quantity of water while the lower element heats the bulk of the water.

Space-heating

Electric space-heating falls into three general groups:

1. High-temperature radiation (known as radiant heating).
2. Low-temperature radiation (floor-warming, block-storage and generally 'black-heat' types).
3. Low-temperature convection ('black-heat' used to promote the circulation of warmed air).

Each type of heating has its own advantages and disadvantages and application.

High-temperature radiation

The advantage of direct (radiant) heating is that it can be switched on and off when and where it is required. High-temperature heating has been described as 'personal warming', in that the heat (from a bright-red element) is radiated direct to a specific area: to be warm a person must be in that area.

Current-using apparatus

The radiant fire is probably the most common direct-heating appliance. It is produced in two forms: (*a*) where the element is a spiral of wire fitted in the grooves of a refractory material (fireclay); (*b*) where the element is a wire wound round a long ceramic cylinder and mounted in front of a reflector of polished metal. A variation of type (*b*) is where the heating element is protected by a tubular sheath of metal or silica and mounted in front of an adjustable angle reflector. These are sometimes called 'infra-red' heaters and are particularly suitable for intermittent heating. They are also the type recommended for bathroom and similar situations where there is the presence of moisture or condensation.

Radiant fires are often used to provide background heating, to supplement the general heating from, say, a central-heating system. These radiant heaters all depend on the element becoming red hot. Ratings vary from 750 W to 3 kW. Heaters can be fixed or portable. Those with two or three separate elements are provided with switches to vary the amount of heat output. All elements should have special guards to prevent direct accidental contact with them (this is a BSI requirement).

Low-temperature radiation

This is an established method of heating. It takes many forms in practice. All are based on a resistive conductor taking an amount of current which will raise its temperature so as to give off radiated heat in small quantities. That is, the element is at 'black heat', which means that the element, though not apparently ON as in the case of the radiator-fire element, still produces heat at a low temperature. Sometimes the elements are embedded between sheets of non-inflammable material such as asbestos, and the sheets fixed to a ceiling. Low-temperature heat is thus radiated downwards into the room. One low-temperature heating system uses heating cables covered by a thermal insulation which forms part of the ceiling structure. Wall-mounted heaters using embedded elements are also used for direct heating; they are often called heater panels. Surface temperatures vary from 49 °C to 66 °C and have loadings from 80 W to 180 W per 100 cm^2 of surface. Surfaces with embedded heating cables reach about 38 °C.

Low-temperature convection heating

This uses the 'black-heat' principle but instead of producing radiated heat, the air surrounding the element is warmed to create convection currents in the air. Basically the convection heater is constructed to draw cold air in from an inlet at the bottom; the rising air is heated by the element and escapes through an outlet at the top. Convection heaters are used for background heating. One example is the low-temperature tubular heater; air in the vicinity of the heater is warmed and so rises, its place being taken by colder air. This establishes a convection current of warm air which is distributed round the room to raise the overall temperature. This type of heater is often fitted with a thermostat situated in a suitable position in the room.

The following are a few of the appliances and apparatuses used for general heating at the present time:

Under-carpet heaters. This consists of a carpet underfelt with concealed high-tensile electric resistance wire in a plastic sheath. The sheath is usually proofed against any liquid which might be spilled on the carpet and allowed to seep through to the underlay. It is tough enough to withstand continual rough traffic. Carpet underlay is made in the standard carpet sizes and is loaded at about 30 W per metre. This is a low-temperature application.

Thermal-storage block-heaters. These are now very common appliances used for central heating in domestic and commercial premises. The usual unit consists of an element surrounded by refractory material, such as concrete. The whole is enclosed in a metal case. Each heater unit is designed to take a continuous charge of electricity over a period of twenty-four hours. The current for these heaters is normally switched on during 'off-peak' periods during the day; the tariff is a cheap-rate type. During the times the current is not on, the heat stored by the heater during its charging periods is released slowly, though some types of unit are fitted with fans to give a boost of heat. Block-storage heaters, as they are sometimes called, are connected to a separate circuit which is separately controlled by a time-switch and has a separate meter.

Floor-warming. This method of heating consists of heating cables buried in the floor (or installed so that they can be removed easily). The temperature produced is about 27 °C. Because the flooring is concrete, or some other heat-storing material, the floor-warming systems are a good off-peak heating load. The installation is controlled by a thermostat to maintain a constant temperature in the room.

Liquid-filled radiator heaters. These resemble the familiar pressed-steel water radiators. They are filled with oil, water or a water-based fluid and contain an element usually rated at 3 kW. Some are also fitted with thermostats.

Fan convector heaters. These have an electric motor-driven fan placed behind a heating element. The purpose of the fan is to give an increased rate of air circulation. This type of heater has the advantage that, when the element is off, cool air can be circulated in a room during hot weather. The domestic type of this heater is known as a 'turbo-convector'.

Installation points

Portable appliances must be effectively earthed. Flexible cables should be sufficient to carry the load current. And the rating of the fuse protecting the appliance, particularly if fed from a 13 A socket-outlet, should be correct. Because the flex is loose, it is subject to more hazards than any other part of an installation. Thus, it is liable to get tugged, rubbed, frayed and stepped on. Flexes should be examined regularly for signs of wear; flex or cord grips should hold the cord sheath securely. Flexibles should be no longer than is necessary to take the supply to an appliance. Appliances are usually supplied with a standard length of flex (generally two metres). If there is no nearby socket-outlet, it is better to provide a new point rather than using an extra-long flex.

User-operated heater appliances

These appliances include electric irons, soldering irons, toasters and similar apparatuses which are used to provide heat for special application.

Electric iron. This appliance consists of an element in tape form wound on a mica former and insulated with mica. The element is clamped between two steel plates and an asbestos pad. The asbestos pad is used to limit the upward flow of heat from the element. The heat is generally controlled by a bimetallic thermostat fitted with a cam-operated control to obtain temperatures ranging from about 120 to 204 °C. An indicating lamp is sometimes fitted to show when the heating element is ON.

Soldering irons. These consist of elements of different types (mica-wound and sheath-type) used to produce heat directed to flow towards a copper bit, the shape of which varies according to the type of duty for which the iron is designed. The wattage of the iron also varies with the type of work to be done.

Motor-operated appliances

In its smallest form, the fractional-kilowatt electric motor has an extremely wide application in domestic, commercial and industrial fields. They include drills, food-mixers, washing machines, spin-driers, refrigerators, ventilating units, hair-driers, fans, clocks and so on. One motor used is the ac series or 'universal' type, so called because it runs on both dc and ac alike. Because ac affects commutation (see Chapter 25 'Motors and control gear') there is a lot of sparking at the brushes, even on light loads. The application for this type of motor includes vacuum cleaners, portable drills and so on. The size of the ac series motor is usually limited to 750 W.

For larger loads, such as washing machines, the motor is one of a number of types of single-phase motors, usually capacitor-start (see Chapter 25).

Fractional kW synchronous motors are used in electric clocks and for other applications where a constant speed is required (the speed depends on the frequency of the supply). This type of motor is self-starting when switched direct-on.

Cooking apparatus

Electric cooking is popular both in the home and in hotels and premises where food has to be prepared. There is an extremely wide range of cooker types, which can be seen on a visit to an electricity showroom. Cooker plates include the solid type, in which the heating element is embedded in a refractory material, and the radiant type (sheathed-element). Each plate can be plugged into a socket for easy installation, removal and replacement. Hot cupboards and grills, eye-level controls, oven lighting, are some of the features of the modern electric cooker. Besides standard cookers there are skillets, rotisseries, table cookers and wall grills. One manufacturer has produced a 'disembodied' cooker system in which the normal parts of a cooker (hotplates, oven, grill and so on) are situated at various points in a kitchen within a specified working area, rather than in one unit as is normal practice.

There are many other applications of electricity which fall into the class of appliance or apparatus and which are to be found in the home, office or for factory light personal duty. These include electric blankets, electric rotary irons, automatic oil and solid-fuel boiler feeds, soil-warming, hedge trimming, electric mowers, and power tools for building operations.

General Regulations requirements

Because of the very wide range of portable electrical appliances available, it is not possible to cover all the electrical requirements for safety in their use. However, the following points have a general application.

All heating equipment must have its supply cables taken into the enclosure with additional heat-resisting sleeving used on the conductors (particularly if they are insulated with PVC). All earthing connections must be continuous between the metal-frame enclosure and the CPC in the flex. The flex rating must be appropriate for the equipment in its current-carrying capacity. Equipment which incorporates a switch (not a thermostat) must be fed from a connection outlet with an appropriate fuse rating. If the equipment has no integral switch, the outlet must be both switched and fused.

Water-heater circuits should be provided with double-pole switches, both in the kitchen and at a position close to the immerser. The flex connecting the immerser unit should be of heat-resisting material and properly supported along its length.

Although the use of shaver-supply units is permitted in bathrooms, they should be located well away from a bath or shower to prevent water splashing onto them.

The control for cooking appliances must be located within 2 m of the appliance. This distance also applies where split-level cooking appliances are used. If this is not possible, a separate double-pole control for each unit is required, though they may form part of the same circuit.

Any portable electrical appliance, such as an electrical lawnmower, must be fed from a socket-outlet which incorporates a residual-current device (RCD) and the socket-outlet must be identified with the words 'For equipment outdoors'.

In commercial and industrial premises where portable equipment is used, the Electricity at Work Regulations require that they be tested periodically with the test results recorded. The tests include insulation resistance and earth bonding. If the equipment is of the double-insulated type (no earth required) a flash test at around 5 kV is also required to ensure that the insulated casing of the equipment is in good condition. It should be noted that these tests are required to be carried out by 'competent persons' which would include a qualified electrician.

One of the aspects of the increasing use of electricity in domestic situations is the installation of a shower cubicle in a bedroom. The Wiring Regulations require that socket-outlets must be located at least 2.5 m from the cubicle. Lampholders must be shrouded or located 2.5 m away from the cubicle. The same distance is required when the cubicle is installed near electrical controls and appliances with touchable heating elements.

13 Lighting and power circuits

Lighting circuits

Most lighting circuits comprise several switching arrangements, such as one-way, two-way and intermediate. A typical domestic circuit is derived from a 5 A way in a consumer unit. In larger installations the rating of the circuit protective device can be either 15 A or 16 A. Domestic lighting is largely based on the use of filament lamps, with fluorescent luminaires found in kitchen areas. Commercial installations are based on fluorescent lamps.

The simplest lighting circuit is a one-way, comprising one lamp controlled by a one-way switch; multi-lamp chandeliers are also controlled from the one-way switch. The two-way switching arrangement is used when a room has two points of entry and the luminaire is thus controlled from any one of two positions. The intermediate switching arrangement is used where one or more luminaires are required to be controlled from a number of switch positions, such as in long corridors or areas where there are a number of entry points.

One-way switches are rated at both 5 A and 15 A. They must always be connected in the phase ('live') side of the supply. When being installed, the switches must be positioned so that when the rocker is in the 'down' position, the circuit is energised. Usually the word 'top' is marked on the reverse of the switch plate to ensure the correct mounting position.

The two-way switch has no OFF position. Rather the lamps are switched ON and OFF by the operation of one or other of the switches. Again, these switches are effectively 'single-pole' control devices and so must be connected to the phase side of the supply.

The intermediate switch has two positions and is effectively a change-over switch. It is connected between two two-way switches and so is able to change the direction of the current in the wires which connect the two-way switches together, known as 'strappers', 'strap wires' or 'pass wires'.

Switches are available as one-gang, or multi-gang, the latter being used to allow a number of individual luminaires to be controlled from one location.

In the provision of lighting circuits in a domestic installation it is usual to split the lighting provision into two, with around ten lamps at the most on each circuit. Each lampholder is assessed at the current equivalent of 100 W. Thus ten lamps would take a total of just over 4 A. The reason for having two lighting circuits is to ensure that part of the house has some light should one circuit fail.

In commercial installations it is essential to ensure that the switches are either rated to carry inductive currents (e.g. 'ac rated') or else are derated to half their normal current rating. This is because fluorescent lamp circuits take more current than the simple rating of the lamp. To calculate the total current taken by fluorescent fittings the lamp wattage is multiplied by a factor of 1.8 and the product divided by the supply voltage:

$$I = \frac{\text{total lamp watts} \times 1.8}{\text{voltage}} \text{ amps}$$

Care should be taken that the correct size of conductor is used for lighting circuits, particularly to prevent excessive voltage drop. Every conductor has a certain value of resistance and when a current flows along it a voltage loss occurs ($V_d = I \times R$). The longer the length of run from its supply point (consumer unit or distribution board) the greater will be the volt drop. If the drop is excessive the lamps (particularly filament lamps) will deliver a reduced light output. This tends to be more a problem in commercial instalations than in domestic premises. Usually conductors of 1.5 mm^2 CSA will be found adequate for most circuits.

Most general lighting circuits are wired either

using the 'loop-in' method or by using a joint box. The loop-in method requires a phase terminal in the ceiling rose (which must always be shrouded to prevent inadvertant contact or else a shock will be received). The other terminals in the rose are for the neutral conductor, the switch wire, and the two terminals to which the luminaire is connected. A final terminal (or terminals) is provided for the CPC which must be connected to both the earth terminal in the switch mounting box and the earth terminal in the luminaire accessory. Some lamp-holders only have two terminals and so the CPC must be terminated in a single block connector and not left loose. The loop-in method using a rose with a phase terminal allows the switches to be fed with two-core cable and also allows other lamps to be looped from the rose.

The joint-box method uses a box made from moulded plastic and can incorporate four terminals or else has a block-connector strip fitted by the electrician. From the box are taken the cables for feeding the switch and the lamp and it also allows other circuits to be taken from the same box.

Note that all CPCs contained in either two- or three-core cables must be sleeved with green/ yellow sleeving when being terminated in ceiling roses, joint boxes, switches or lamp accessories.

If one-way switching is involved, the red core must be the switch feed conductor. The other core, coloured black, actually forms the switch wire and, to comply with the Wiring Regulations, must be identified as being on the 'live' side of the circuit and coloured red, yellow or blue depending on the phase from which the installation is being supplied. This identification of the switch wire is often ignored by practising electricians and while it is general practice it must be recognised that it does contravene the Regulations.

In commercial and industrial installations, the 'live' may be derived from the red, yellow or blue phase and its colour should be consistent throughout the circuit.

Power circuits

Two types of circuits are used to feed socket-outlet units and fused connection units: radial and ring.

The radial circuit is derived from a 20 A way in a distribution board or consumer unit and is intended to serve a floor area not exceeding 20 m^2. If the floor area to be served is not more than 50 m^2 the rating of the protective device is 30 A or 32 A. Kitchens in domestic premises (where a large number of socket-outlets are often needed for electrical appliances) are often fed by radial circuits.

The ring circuit is derived from a 30 A or 32 A way and has its conductors terminated at its point of origin in the distribution board or consumer unit. An unlimited number of socket-outlets may be connected to the circuit provided that the floor area served does not exceed 100 m^2 in domestic installations. As most modern domestic premises have a floor area in excess of this figure, two ring circuits are usually provided (or one ring and one radial circuit for the kitchen). When more than one ring circuit is installed in the same premises, the socket-outlets should be reasonably shared between the circuits to balance the loading.

Spurs can be taken off a ring circuit. If they are of the non-fused type, the number must not exceed the number of socket-outlets (and any stationary equipment) connected directly in the ring. Double or two-gang units are regarded as two separate outlets. Connection units can be either fused or switched and fused, with the latter being used for equipment which has no direct control switch. A fused connection unit can be used to feed devices which take very little current, such as a fan, a clock or a lamp.

When installing a ring circuit, the spurs can be fed from socket-outlets connected directly in the ring, from the origin of the circuit or from a joint box (with terminals rated for 30 A) connected in the ring and left for future extensions to a building (e.g. a conservatory or extra bedroom).

Three types of cable are recognised for use with ring and radial circuits: PVC-sheathed, MI cable and copper-clad aluminium, PVC-insulated. For ring circuits the respective csa involved are 2.5, 1.5 and 4 mm^2.

Regulations summary

Socket-outlets (except for shaver units) are not permitted in bathrooms. Fixed appliances fed from connection units must have an appropriate provision in the unit (either fused or switched and fused). Appliances connected to fused connection units must have a rating of 3 kW maximum. A two-gang socket-outlet unit is regarded as two separate units. Any socket-outlet used for supplying electrical equipment intended for outdoor use must incorporate a residual current device (RCD) with a trip current of 30 mA; the socket-outlet must be next to a warning label reading 'For equipment outdoors'. In industrial or commercial premises, adjacent socket-outlets must be connected to the same phase of the supply.

All socket-outlets must be protected by a device (fuse or miniature circuit-breaker, MCB) which will operate within 0.4 second should a fault occur. The cable sizes recommended for both ring and radial circuits assume normal ambient temperatures and that the conductors are not bunched. Higher operating temperatures, bunching, contact with thermal insulation material and the use of a semi-enclosed fuse (BS 3036) may mean an increase in the csa of the conductors.

Cooker circuits

These are derived from (usually) a 30 A or 32 A way, but can be higher depending on the kW rating of the cooker. The control units need not have a socket-outlet incorporated in them, but if one is provided the protective device must be able to disconnect the circuit in the event of a fault within 0.4 second (it would be 5 seconds otherwise). The control unit must be located within 2 m of the appliance (this also applies to 'split-level' cooking appliances).

Water-heater circuits

Generally derived from a 15 A or 16 A way, the circuit must incorporate a double-pole switch (usually of 20 A rating), with an additional switch recommended in close proximity to the immerser unit which should be connected by using heat-resisting flexible cable or cord.

Voltage drop in circuits

If any conductor carries a current there will be a loss of voltage between both ends of the conductor. This loss or 'volt drop' is $V = I \times R$, where V is the volt drop, I is the current in the conductor and R is the conductor resistance. The effect of serious volt drop is to reduce the effective performance of lamps, heaters and other electrical devices. To limit volt drop in any circuit, the IEE Regulations impose a maximum drop of 4 per cent of the nominal circuit voltage. Thus, on a 240 V supply, the drop between the consumer's terminals and the farthest end of any circuit in the installation is 9.6 V. It is the concern of the installation designer, and often the electrician, to choose a size of circuit conductor for a particular load so that this figure is not exceeded. The Current-rating Tables in the Regulations indicate the volt drop (in millivolts [mV]) when a current of 1 A flows through a 1 metre length of a particular cable. This mV figure is then multiplied by the actual load current in the cable and also by the length of cable run, to arrive at the total volt drop. If, for a particular size of cable, the total volt drop exceeds 9.6 V (for a 240 V supply), the next size of cable is chosen in turn until the final volt drop is less than 9.6 V. Other factors are, however, required to be taken into account when choosing the initial size of cable, such as the cable's ambient temperature, the cable's installation condition (e.g. bunched with other cables), the circuit protection (e.g. semi-enclosed fuse or MCB), and whether the cable is in contact with thermal insulation material. The IEE Regulations require that the choice of a cable to feed a particular circuit must have regard for a number of factors, and not just the circuit current.

The Regulations require that the method used to choose the correct size of a conductor be based on the rating of the protective overcurrent device. All factors which affect the rating of the cable in its installed condition are applied as divisors to the rating of the protective device, as the following examples show. The process involved in working out the correct cable size, and the final volt drop, is as follows:

1. First find the load current of the circuit (I_B).
2. Determine the correction factor for the ambient temperature in which the cable is to

be installed (the highest temperature is always taken).

3. Determine the correction factor for grouping C_g, if the cable is run with others.
4. Determine the correction factor C_i if the cable is in contact with, or surrounded by, thermal insulation material. Two factors are given: 0.75 if only one side of the cable is in contact with the material (e.g. cable clipped to a joist) and 0.5 if the material completely surrounds the cable.
5. Select the rating of the overcurrent device. If this offers what used to be called 'close protection', e.g. by an MCB, the factor is 1.0. If, however, the device is a semi-enclosed fuse, the factor is reduced to 0.725. In any case, the rating of the device must equal the circuit load current.
6. Determine the size of the circuit conductor, by calculating the desired current rating.
7. Check that the volt drop does not exceed the maximum permissible allowed.

If I_z represents the current rating of the conductor, and I_n the rating of the protective device, then

$$I_z = \frac{I_n}{C_g \times C_a \times C_i \times 0.725}$$

where C_g is the correction factor for grouping, C_a is the factor for ambient temperature, C_i is the factor for the thermal insulation, if applicable, and 0.725 is the factor for the overcurrent protective device, if applicable.

Example 1

A 240 V single-phase load of 22 A is supplied from a distribution board located at a distance of 38 m. The cables are to be single-core, with 85 °C insulation and are to be enclosed in metallic conduit. The ambient temperature is taken as 60 °C. Protection is by MCB.

Maximum permissible volt drop (MPVD) = 4 per cent of 240 V = 9.6 V

Load current (I_B) = 22 A

Rating of protective device (I_n) = 30 A (see Fig. 6, Appendix 3)

Current rating of conductor (I_z)

$$= \frac{I_n}{\text{Correction factor for 60 °C}} \text{ amps}$$

Correction factor for 60 °C = 0.67
(from Table 4C1)

$$\therefore I_z = \frac{30}{0.67} = 44.7 \text{ A}$$

The required rating of the conductor (from Table 4F1A) is 6 mm^2

The total volt drop is

$$\text{VD/A/m} = \frac{I_B \times \text{length of run}}{1000}$$

$$= \frac{7.7 \times 22 \times 38}{1000} = 6.44 \text{ V}$$

Thus the conductor size of 6 mm^2 is suitable for the load in the conditions specified.

Example 2

A 230 V single-phase load of 15 kW operates at a power factor of 0.85 and is fed from a distribution board located 40 m away by a two-core paper-insulated and armoured cable with aluminium conductors. The cable is clipped direct to a cable tray which also carries two other similar cables. The ambient temperature is 45 °C and the circuit is protected by a fuse to BS 88, Part 2.

MPVD = 4 per cent of 230 V = 9.2 V

Load current (I_B)

$$= \frac{\text{W}}{\text{V} \times \text{p.f.}} = \frac{15\,000}{230 \times 0.85} = 76.7 \text{ A}$$

From Fig. 3B, Appendix 3, fuse rating = 100 A (I_n)

Correction factors:

A — Grouping (Table 4B1) = 0.81
B — Ambient temperature (Table 4C1) = 0.84

Required cable size (I_z)

$$= \frac{100}{0.81 \times 0.84} = 147 \text{ A}$$

From Table 4D4A/, size of cable is 35 mm^2

$$\text{Final volt drop} = \frac{1.25 \times 76.7 \times 40}{1000} = 3.83 \text{ V}$$

Thus the calculated cable size of 35 mm^2 will satisfy the conditions of the load.

Example 3

A 240 V, 4 kW fixed resistive load is to be fed by a PVC-insulated and sheathed cable. The installation condition involves the cable being run with four other similar cables surrounded by glass-fibre thermal insulation. The ambient temperature is assessed as 40 °C. Protection is by a semi-enclosed fuse to BS 3036. The length of run is 18 m.

MPVD = 4 per cent of 240 V = 9.6 V

$$\text{Load current } (I_B) = \frac{4000}{240} = 16.7 \text{ A}$$

Rating of fuse (I_n) (from Fig. 2B, App. 3) = 20 A

Correction factors:

A — Grouping for five cables (Table 4B1, App. 4) = 0.6
B — Ambient temperature of 40 °C (Table 4C2, App. 4) for cables protected by semi-enclosed fuses = 0.94
C — Thermal insulation = 0.5
D — An additional factor is involved because semi-enclosed fuses are provided = 0.725

Rating of cable (I_z)

$$= \frac{20}{0.6 \times 0.94 \times 0.5 \times 0.725} = 98 \text{ A}$$

Size of cable (Table 4D2A) Installation methods ('enclosed') Table 4B1 is 50 mm^2

$$\text{Volt drop} = \frac{0.94 \times 16.7 \times 18}{1000} = 0.28 \text{ V}$$

Thus 50 mm^2 is the suitable size of cable for this load.

Example 4

A fixed load of 56 A is carried by two single-core MICS cables with bare sheath and installed so as not to be exposed to touch. The terminations used are rated for 105 °C. The length of run from the 240 V single-phase supply point is 35 m. The ambient temperature is taken as 80 °C. The cables are run-clipped direct to a cable tray with three other single-phase circuits run in MICS cables. The cables are protected by semi-enclosed fuses to BS 3036.

MPVD = 4 per cent of 240 V = 9.6 V

Load current (I_B) = 56 A

Rating of fuse (I_n) (from Fig. 2B, App. 3) = 100 A

Correction factors:

A — Grouping: No factor need be applied in this case
B — Ambient temperature (Table 4C2) for 80 °C = 0.77
C — Factor for the use of BS 3036 fuses is 0.725

$$\text{Cable rating required } (I_z) = \frac{100}{0.77 \times 0.725} = 179 \text{ A}$$

From Table 4J2A the cable size is 35 mm^2

$$\text{Total volt drop} = \frac{1.3 \times 56 \times 35}{1000} = 2.55 \text{ V}$$

Thus 35 mm^2 is satisfactory for this circuit.

14 Electrical safety

The concern for electrical safety has a long-standing place in the Statutory legislation of the country, starting with the 1901 Factories and Workshop Act which was used as a starting-point to bring in the 1908 Electricity Regulations. But even before these dates, the Society of Telegraph Engineers and Electricians in 1882 published the Rules and Regulations for electric wiring. The Society is now the more august Institution of Electrical Engineers. Over the decades, since the turn of the century, the legislation has been revised to include the increasing use of electricity in many situations other than factories. In 1974 the Health and Safety at Work Act came into force with the electrical aspects of that Act, stated and inferred, being developed into the Electricity at Work Regulations, 1989. This Act became law in April 1990. The requirements of the Act bring in some 20 million people who were not otherwise protected against possible risk of shock and burns resulting from contact with live parts of equipment and wiring systems. The Act does not apply to domestic premises (except where work is being carried out by an electrical contractor and his employees). It does include hospitals, construction sites, schools, colleges, commercial premises, factories and mines and quarries.

The Electricity at Work Regulations (EAWR) take in all electrical equipment such as switchgear, distribution boards, electrial accessories, portable tools and cables.

The construction industry in this country still contributes to an unsatisfactory safety record with, in 1990, 151 people killed and 4,200 people suffering serious injuries — all sustained on construction sites. The number of people who are killed by electrocution is well under 100 each year, which includes fatalities involving electrical equipment in the home, such as space-heating, cookers, blankets and bedwarmers. In 1990 17 people were killed in electrocution accidents in the home with several of the fatalities traced to faulty 13 A plugs alone.

The causes of accidents and fatalities are many and varied and range from faulty wiring, the misuse or abuse of electrical equipment, DIY activities, faulty plugs and socket-outlets, damaged connectors, leads and extension leads. Some fatalities have been attributed to inadequate earth continuity.

The figures already given above only record deaths from electrocution. Other fatalities occur as a result of fires started in electrical equipment and wiring systems, with much of the latter involving old installations which have either been neglected or else have imposed on them extra loads so that the inadequacy of the conductors shows up in heat and fire. Even though the Wiring Regulations recommend that installations be tested at regular intervals (five years for a domestic installation, for example) there is no obligation on the owner of a premises to have the installation tested and inspected. The reluctance is mainly due to the fact that a test and inspection costs money even though the 'peace of mind' would be worth the cost. The normal domestic installation user regards the wiring as something which, if it works, need not be looked after.

Commercial and industrial installations tend to be subject to regular inspection, testing and maintenance, either on a maintenance conract basis or by employing qualified electricians. Thus, because these premises come under Statutory regulations, installations tend to be less liable to become dangerous. The EAWR will no doubt ensure that the annual rate of fatalities will be reduced still further. The problem of domestic premises will remain until Statutory legislation is introduced to compel domestic installation owners to make sure their electrial provisions are safe to use.

The Wiring Regulations are greatly concerned with the protection of the user of electrical instal-

Figure 14.1. Fires attributed to electrical causes.

Occupancy	*Wire and cable*	*Lighting*	*Space heating*	*Cooking appliances*	*Other*	*Total*
Industry (inc. construction)	451	77	134	55	1,122	1,839
Shops	251	190	58	63	193	755
Agricultural premises	192	35	94	1	40	362
Hotels, hostels, boarding houses	85	39	57	217	117	515
Restaurants, clubs, public houses	205	90	16	259	140	710
Places of public entertainment	47	22	9	15	36	129
Hospitals	42	78	23	91	234	468
Schools	53	31	34	36	50	204
Other	820	347	243	208	1,042	2,660
Total	2,146	909	668	945	2,974	7,642

lations, whether in the home or in industry. The Regulations are quite specific about the factors which have to be taken into account when an installation is being designed and then installed. The philosophy is that if an installation is well designed and care is taken to ensure that approved materials and good workmanship are involved, the end product reduces the risk of fire and shock.

The safety provision in any electrical installation is (*a*) to prevent electric shock and (*b*) to prevent the occurrence of fires due to electrical causes (see Fig. 14.19).

Full explanations of these preventive measures are detailed in Chapter 15 'Protection' and in Chapter 16 'Earthing'. However, a few words about the human element in an electrical installation will be useful to give a background knowledge of how electrical accidents can arise.

For an electric shock, it is necessary for the human body to be in contact with two objects of unequal potentials in such a way that the body forms part of an electrical circuit in which current will flow. The amount of current flowing through the body will then decide on how serious the accident will become. The currents which cause the following conditions in the human body are:

1–3 milliamps*: This is known as the 'threshold of perception' when a slight tingling sensation is felt.

10–15 milliamps: At this value of current the muscles begin to tighten and it becomes difficult to release any object held (say, a live conductor).

25–30 milliamps: At this current the muscles are really tight and the person has absolutely no control over them. This is the first dangerous state.

over 50 milliamps: At this current fibrillation of the heart occurs, which is generally lethal if immediate specialist attention is not given.

It is thus seen that quite small currents can be fatal, particularly if the person in contact with a live object has a weak heart. As a point of interest, the current taken by a 240 V, 15 W filament lamp is 62 milliamps.

The most common method used today for the protection of human beings against the risk of electric shock is either (*a*) the use of insulation (screening live parts, and keeping live parts out of reach) or (*b*) ensuring, by means of earthing, that any metal in an electrical installation, other than the conductor, is prevented from becoming electrically charged. Earthing basically provides a path of low resistance to earth for any current which results from a fault between a live conductor and earthed metal.

* A milliamp is one-thousandth of an amp.

The general mass of earth has always been regarded as a means of getting rid of unwanted currents; charges of electricity could be dissipated by conducting them to an electrode driven into the ground. A lighting discharge to earth illustrates this basic concept of the earth as being a large 'drain' for electricity. Thus, every electrical installation which has metalwork associated with it (either the wiring system, accessories or the appliances used) is connected to earth. Basically, this means that if, say, the framework of an electric fire becomes 'live', the resultant current will, if the frame is earthed, flow through the frame, its associated circuit-protective conductor, and thence to the general mass of earth. Earthing metalwork by means of a bonding conductor means that all the metalwork will be at earth potential; or, no difference in potential can exist. And because a current will not flow unless there is a difference in potential, then the installation is said to be safe from the risk of electric shock. Reading through Chapter 16 will indicate that the subject of earthing is an important one and merits close attention by anyone who aims to work with electricity.

Effective use of insulation is another method of ensuring that the amount of metalwork in an electrical installation which could become live is reduced to a minimum. The term 'double-insulated' means that not only are the live parts of an appliance insulated, but that the general construction is of some insulating material. A hair-dryer and an electric shaver are two items which fall into this category.

Though the shock risk in every electrical installation is something with which every electrician must concern himself, there is also the increase in the number of fires caused, not only by faults in wiring, but also by defects in appliances. In order to start a fire there must either be sustained heat or an electric spark of some kind. Sustained heating effects are often to be found in overloaded conductors, bad connections, loose-fitting contacts and so on. If the contacts of a switch are really bad, then arcing will occur which could start a fire in some nearby combustible material, such as blockboard, chipboard, sawdust and the like. The purpose of a fuse is to cut off the faulty circuit in the event of an excessive current flowing in the circuit. But fuse-protection (as indicated in Chapter 15) is not always a guarantee that the circuit is safe from fire risk. The wrong size of fuse, for instance (e.g. 15 A wire instead of 5 A wire) will render the circuit dangerous.

Fires can also be caused by an earth-leakage current causing arcing between live metalwork and, say, a gas pipe. Again, fuses are not always of use in the protection of a circuit against the occurrence of fire. Residual-current devices (RCD) (see Chapter 16) are often used instead of fuses to detect small fault currents and to isolate the faulty circuit from the supply.

To ensure a high degree of safety from shock-risk and fire-risk, it is thus important that every electrical installation be tested and inspected not only when it is new but at periodic intervals during its working life. Many electrical installations today are anything up to fifty years old. And often they have been extended and altered to such an extent that the original safety factors have been reduced to a point where amazement is expressed on why 'the place hasn't gone up in flames before this'. Insulation, used as it is to prevent electricity from appearing where it is not wanted, often deteriorates with age. Old, hard and brittle insulation may, of course, give no trouble if left undisturbed and is in a dry situation. But the danger of shock — and fire-risk — is ever present,. for the cables may at some time be moved by electricians, plumbers, gas-fitters and builders.

It is a recommendation of the IEE Regulations that every domestic installation be tested at intervals of five years or less. The Completion and Inspection Certificates in the IEE Regulations show the details required in every inspection. And not only should the electrical installation be tested, but all the current-using appliances and apparatus used by the consumer. The values of insulation resistance, earth impedance and so on are indicated in Chapter 18.

The following are some of the points which the inspecting electrician should look for:

1. Flexible cables not secure at plugs.
2. Frayed cables.
3. Cables without mechanical protection.

4. Use of unearthed metalwork.
5. Circuits over-fused.
6. Poor or broken earth connections, and especially signs of corrosion.
7. Unguarded elements of radiant fires.
8. Unauthorised additions to final circuits resulting in overloaded circuit cables.
9. Unprotected or unearthed socket-outlets.
10. Appliances with earthing requirements being supplied from two-pin BC adaptors.
11. Bell-wire (with extra-low voltage insulation) used to carry mains voltages.
12. Use of portable heating appliances in bathrooms.
13. Broken connectors, such as plugs.
14. Signs of heating at socket-outlet contacts.

In the normal course of his/her work in installing wiring systems, electrical accessories and equipment, the electrician is building into the installation all the fundamental requirements for electrical safety. In a practical form, these requirements include:

1. Ensuring that all conductors are sufficient in csa for the design load current of circuits.
2. All equipment, wiring systems and accessories must be appropriate to the working conditions.
3. All circuits are protected against overcurrent using devices (fuses, MCBs) which have ratings appropriate to the current-carrying capacity of the conductors.
4. All exposed conductive parts (i.e. all metalwork associated with the electrical installation) are connected together by means of CPCs.
5. All extraneous conductive parts (i.e. metalwork not directly associated with the installation, such as radiators, water pipes, structural steelwork and services fed by metal pipes) are bonded together by means of main bonding conductors and supplementary bonding conductors, and that these conductors are taken to the installation main earth terminal.
6. All control and overcurrent protective devices are installed in the phase conductor.
7. All electrical equipment have the means for their control and isolation.
8. All joints and connections must be mechanically secure and electrically continuous and be accessible at all times.
9. No additions to existing installations should be made unless the existing conductors are sufficient in size to carry the extra loading.
10. All electrical conductors have to be installed with adequate protection against physical damage and be suitably insulated for the circuit voltage at which they are to operate.
11. In situations where a fault current to earth is not sufficient to operate an overcurrent device, an RCD must be installed.
12. All electrical equipment intended for use outside the equipotential zone (i.e. the premises where all metalwork is both bonded and earthed) must be fed from socket-outlets incorporating an RCD.
13. The detailed inspection and testing of installations before they are connected to a mains supply, and at regular intervals thereafter.

Note: A recent survey by the Royal Society for the Prevention of Accidents assessed the safety of 13 A plugs in 20 000 homes. Some 50 000 plugs were found to be faulty with 4 per cent declared 'Positively dangerous'. Faults included: live and neutral wires reversed; earth terminal not connected; terminal screws loose; cord clamp missing or not used; insulation damaged or stripped back too far; two appliances connected to the same plug.

15 Protection

In electrical work the term 'protection' is applied to precautions to prevent damage to wiring systems and equipment, but also takes in more specific precautions against the occurrence of fire due to overcurrents flowing in circuits, and electric shock risks to human beings as a result, usually, of earth-leakage currents appearing in metalwork not directly associated with an electrical installation, such as hot and cold water pipes.

The initial design of any installation must take into account the potential effects on wiring systems and equipment of environmental and working conditions. BS 5490 is a British Standard concerned with protection against mechanical, or physical, damage and gives full details of the Index of Protection Code to which all electrical equipment must conform. The Code is based on a numbering system with each number indicating the degree of protection offered.

The first characteristic numeral indicates the protection level offered to persons against contact with live or moving parts inside an enclosure and also the protection of the enclosure itself against the ingress of solid bodies, such as dust particles. The numbers range from 0 (no protection of equipment against the ingress of solid bodies and no protection against contact with live or moving parts) to 6 (complete protection).

The second characteristic numeral indicates the degree of protection of equipment against the ingress of liquid and ranges from 0 to 8. Thus an equipment with IP44 means that there is protection against objects of a thickness greater than 1.0 mm and against liquid splashed from any direction.

Mechanical damage

This term includes damage done to wiring systems, accessories and equipment by impact, vibration and collision, and damage due to corrosion (see later in this chapter). Typical examples of prevention include single-core conductors in conduit and trunking, the use of steel enclosures in industrial situations, the proper supporting of cables, the minimum bending radius for cables, the use of armoured cables when they are installed underground, and the supports required for conductors in a vertical run of conduit and trunking.

Some types of installation present greater risks of damage to equipment and cables than others, for example on a building or construction site and in a busy workshop. In general, the working conditions should be assessed at the design stage of an installation and, if they have not been foreseen, perhaps due to a change of activity in a particular area, further work may be needed to meet the new working conditions.

Electrical fires are caused by (*a*) a fault, defect or omission in the wiring, (*b*) faults or defects in appliances and (*c*) mal-operation or abuse of the electrical circuit (e.g. overloading). The electrical proportion of fire causation today is around the 20 per cent mark. The majority of installation fires are the result of insulation damage, that is, electrical faults accounting for nearly three-quarters of cables and flex fires. Another aspect of protection against the risk of fire is that many installations must be fireproof or flameproof. The definition of a flameproof unit is a device with an enclosure so designed and constructed that it will withstand an internal explosion of the particular gas for which it is certified, and also prevent any spark or flame from that explosion leaking out of the enclosure and igniting the surrounding atmosphere. In general, this protection is effected by wide-machined flanges which damp or otherwise quench the flame in its passage across the metal, but at the same time allows the pressure generated by the explosion to be dissipated.

One important requirement in installations is the need to make good holes in floors, walls and ceilings for the passage of cables, conduit, trunking and ducts by using incombustible materials to prevent the spread of fire. In particular, the use of fire barriers is required in trunking.

It was not until some years after the First World War that it was realised there was a growing need for special measures where electrical energy was used in inflammable situations. Precautions were usually limited to the use of well-glass lighting fittings. Though equipment for use in mines was certified as flameproof, it was not common to find industrial gear designed specially to work with inflammable gases, vapours, solvents and dusts. With progress, based on the results of research and experience, a class of industrial flameproof gear eventually made its appearance and is now accepted for use in all hazardous areas.

There are two types of flameproof apparatus: (*a*) mining gear, which is used solely with armoured cable or special flexibles; and (*b*) industrial gear, which may be used with solid-drawn steel conduit, MIMS cables, aluminium-sheathed cables or armoured cables. Mining gear is known as 'Group I' gear and comes into contact with only one fire hazard: firedamp or methane. Industrial gear, on the other hand, may well be installed in situations where a wide range of explosive gases and liquids are present. Three types of industrial hazards are to be found: explosive gases and vapours — inflammable liquids — and explosive dusts. The first two hazards are covered by what is called 'Group II' and 'Group III' apparatus. Explosive dusts may be of either metallic or organic origin. Of the former, magnesium, aluminium, silicon, zinc and ferro-manganese are hazards which can be minimised by the installation of flameproof apparatus; the flanges of which are well greased before assembly. The appropriate British Standard Code of Practice is BS 5345 *Electrical Apparatus and Associated Equipment for Use in Explosive Atmospheres of Gas or Vapour, other than Mining Applications*.

All equipment certified as 'flameproof' carries a small outline of a crown with the letters Ex inside it. The equipment consists of two or more compartments. Each is separated from the other by integral barriers which have insulated studs mounted therein to accommodate the electrical connection. Where weight is of importance, aluminium alloy is permitted. All glassware is of the toughened variety to provide additional strength. The glass is fitted to the apparatus with a special cement. Certain types of gear, such as distribution boards, are provided with their own integral isolating switches, so that the replacement of fuses, maintenance, and so on, cannot be carried out while a circuit is live.

All conduit installations for hazardous areas must be carried out in solid-drawn 'Class B', with certified draw-boxes, and accessories. Couplers are to be of the flameproof type with a minimum thread length of 50 mm. All screwed joints, whether entering into switchgear, junction boxes or couplers, must be secured with a standard heavy locknut. This is done to ensure a tight and vibration-proof joint which will not slacken during the life of the installation, and thus impair both continuity and flameproofness. The length of the thread on the conduit must be the same as the fitting plus sufficient for the locknut. Because of the exposed threads, running couplers are not recommended. Specially designed unions are manufactured which are flameproof and are designed to connect two conduits together or for securing conduit to an internally threaded entry.

Conduits of 20 and 25 mm can enter directly into a flameproof enclosure. Where exposed terminals are fitted, conduits above 25 mm must be sealed at the point of entry with compound. Where a conduit installation is subject to condensation, say, where it passes from an atmosphere containing one type of vapour to another, the system must be sectionalised to prevent the propagation of either condensated moisture or gas. Conduit stopper boxes, with two, three or four entries, must be used. They have a splayed, plugged filling spout in the cover so that the interior can be completely filled with compound.

When flexible, metal-sheathed or armoured cables are installed, certified cable glands must be used. Where paper-insulated cables are used, or in a situation where sealing is necessary, a cable-sealing box must be used, which has to be filled completely with compound.

The following are among the important installation points to be observed when installing flameproof systems and equipment. Flanges should be greased to prevent rusting. Special care is needed with aluminium-alloy flanges as the metal is ductile and is easily bent out of shape. All external bolts are made from special steel and have shrouded heads to prevent unauthorised interference; bolts of another type should not be fitted as replacements. Though toughened glass is comparatively strong, it will not stand up to very rough treatment; a faulty glass will disintegrate easily when broken. Protective guards must always be in place. Conduit joints should always be painted over with a suitable paint to prevent rusting. Because earthing is of prime importance in a flameproof installation, it is essential to ensure that the resistance of the joints in a conduit installation, or in cable sheaths, is such as to prevent heating or a rise in voltage from the passage of a fault-current. Remember that standard flameproof gear is not necessarily weather-proof, and should be shielded in some way from rain or other excessive moisture.

Being essentially a closed installation, a flameproof conduit system may suffer from condensation. Stopper boxes prevent the passage of moisture from one section to another. Draining of condensate from an installation should be carried out only by an authorised person. Alterations or modifications must never be made to certified flameproof gear. Because flexible metallic tubing is not recognised as flameproof, cables to movable motors (e.g. on slide-rails) should be of the armoured flexible cable type, with suitable cable-sealing boxes fitted at both ends. It is necessary to ensure that, as far as possible, contact between flameproof apparatus, conduit, or cables, and pipework carrying inflammable liquids should be avoided. If separation is not possible, the two should be effectively bonded together. When maintaining equipment in hazardous areas, care should be taken to ensure that circuits are dead before removing covers to gain access to terminals. Because flexible cables are a potential source of danger, they should be inspected frequently. All the equipment should be inspected and examined for mechanical faults, cracked glasses, deterioration of well-glass cement, slackened conduit joints and corrosion. Electrical tests should be carried out at regular intervals.

Corrosion

Wherever metal is used, there is often the attendant problem of corrosion and its prevention. There are two necessary conditions for corrosion: (*a*) a susceptible metal and (*b*) a corrosive environment. Nearly all of the common metals corrode under most natural conditions. Little or no specific approach was made to the study of corrosion until the early years of the nineteenth century. Then it was discovered that corrosion was a natural electrochemical process or reaction by which a metal reverts in the presence of moisture to a more stable form usually of the type in which it is found in nature. It was Humphry Davy who suggested that protection against corrosion could result if the electrical condition of a metal and its surroundings were changed.

Corrosion is normally caused by the flow of direct electrical currents which may be self-generated or imposed from an external source (e.g. an earth-leakage fault-current). Where direct current flows from a buried or submerged metal structure into the surrounding electrolyte (the sea or soil), no corrosion takes place. It is an interesting fact to record that where a pipe is buried in the soil there is a 'natural' potential of from -0.3 V to -0.6 V between the pipe and the soil. In electrical installations, precautions against the occurrence of corrosion include:

(*a*) The prevention of contact between two dissimilar metals (e.g. copper and aluminium).

(*b*) The prohibition of soldering fluxes which remain acidic or corrosive at the completion of a soldering operation (e.g. cable joint).

(*c*) The protection of cables, wiring systems and equipment against the corrosive action of water, oil and dampness, unless they are suitably designed to withstand these conditions.

(*d*) The protection of metal sheaths of cables and metal conduit fittings where they come into contact with lime, cement and plaster and certain hard woods (e.g. oak and beech).

(*e*) The use of bituminised paints and PVC oversheathing on metallic surfaces liable to corrosion in service.

Dampness can affect conduit systems both on the inside and externally. With enamel finishes, it is important that the enamel is preserved as intact as possible, particularly at the thread entry to fittings. Also, the breaking of the galvanising finishing on galvanised conduit presents a great risk of rusting simply because this type of conduit was specified to cope with damp or wet working conditions. Thus any breaks in the finish must be repaired with the use of a suitable paint to prevent rusting.

Internal corrosion can occur in situations where the ambient temperature tends to fluctuate. Condensation thus occurs, even in what would otherwise be dry situations, and if the resulting condensate is not allowed to drain away out of the conduit run a build-up can occur. To deal with this problem, the drainage points are recommended in the form of conduit boxes either with holes drilled to allow condensate to drip out or else, say, using a tee box with the T-outlet plugged with a plug which can be removed at intervals.

Special care is needed in the choice of materials for clips and other fittings for bare aluminium-sheathed cables, and for aluminium conduit, because aluminium is not particularly stable in damp situations and especially when in contact with other metals. For instance, fixing an aluminium bulkhead luminaire with brass screws to an external wall can set up an electrolytic action between the fitting and the screws. Chromium-plated screws would be better in this situation.

While copper is fairly resistant to corrosion, there are situations in which the material will corrode. This is why MI copper-sheathed cables are provided with PVC-sheaths and clips are also covered with PVC.

Protective enclosures for electrical apparatus

It is often required of a piece of electrical equipment that it be fitted in an enclosure which will protect it from physical damage because of some installation condition. Protection against mechanical damage, fire and corrosion has already been mentioned. The following are some other types of protection, usually in the form of an enclosure which is designated as splashproof, dustproof, and so on, when so constructed, protected or treated that satisfactory operation of the enclosed equipment is not interfered with, if subjected to the specified condition.

(*a*) *Screen-protected.* Enclosure has the openinngs covered by screens of wire-mesh, expanded metal or other perforated material.

(*b*) *Totally enclosed.* Enclosure with no openings for ventilation, but not necessarily airtight.

(*c*) *Waterproof.* The enclosure will exclude water under prescribed conditions which include a limited period of submersion.

(*d*) *Drip-proof.* Enclosure with openings so protected that liquid or solid particles falling on it cannot enter to the enclosed apparatus.

(*e*) *Weatherproof.* The enclosure is able to withstand exposure to sun, rain, mist, snow and airborne particles.

The British Standard 5000 *Rotating Electrical Applications* indicates the forms of enclosures for electric motors, and suggests the factors to be considered when installing a machine in a particular environment.

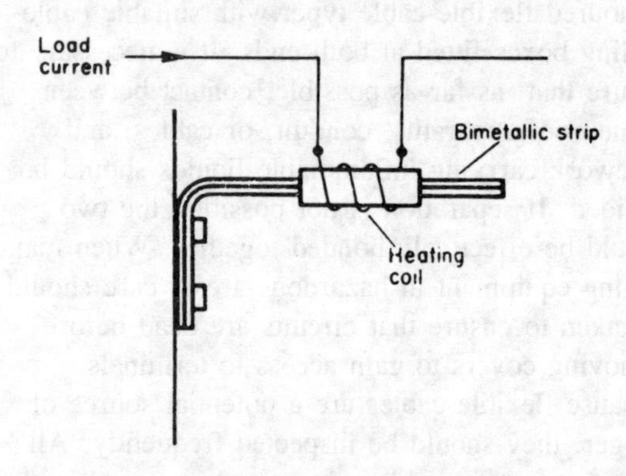

Figure 15.1. Typical bimetallic device.

Under-voltage

This is an electrical protection required by Regulation 552-4, and is a provision in the circuit of an electric motor to prevent automatic restarting after a stoppage of the motor due either to an excessive drop in the supply voltage, or a complete failure of the supply, where unexpected restarting of the motor might cause injury to an operator. These devices are found in dc motor starters (No-volt releases). In ac contactor starters failure of the supply stops the motor.

Protective relay equipment

The protective relay is basically a series-connected, direct-acting, over-load trip coil. It is electromagnetic in operation and consists of either an electromagnet and armature, or a solenoid and a central plunger. The coil of the electromagnet (or the solenoid) consists of a few turns of conductor connected in series with the main circuit. The armature (or plunger) is arranged to operate the associated circuit-breaker trip mechanism. The operating current value of such a device is usually adjustable by varying either the magnetic gap of the electromagnet (or solenoid plunger) or a restraining spring. A time-lag or time-delay can be introduced in the relay circuit by using an air or oil dashpot attached to the movement of the trip device, so that a machine or transformer can take an overload for a short time without serious overheating.

The thermal relay is suitable for giving overload protection to motors and transformers. The principle of operation is based on the well-known action of bimetallic expansion in which two dissimilar metals, when coupled together in the form of a strip or spring, are arranged to give a deflection under the influence of heat. The amount of deflection increases with the amount of heat, until the point is reached where the operation of the relay results due to the closing of its contacts. If a proportion of the load current is arranged to pass through the bimetal strip or spring, the resultant heat will cause the latter to deflect. By predetermining the amount of deflection required to close the relay contacts, the circuit-breaker can be tripped at any value of the load current it carries. This type of relay has a relatively slow action, because of the time taken to heat up the bimetal element. But this fact is often used as an advantage where overload protection is concerned, since it introduces a suitable time lag. Therefore, it will not operate on momentary or transient overloads such as occur during the starting of a motor, but will do so if the overload is sustained for a predetermined length of time.

Overcurrent

Overcurrent or excess current is the result of either an overload or a short-circuit. Overloading occurs when an extra load is taken from the supply. This load, being connected in parallel with the existing load in a circuit, decreases the overall resistance of the circuit with an attendant rise in the current flowing in the circuit. This increased current will have an immediate effect on the circuit cables: they will begin to heat up. If the overload is sustained the result will be an accelerated deterioration of the cable insulation and the eventual breakdown of it to cause an electrical fault or fire. It is obvious, then, that some means of protection must be incorporated in a circuit to prevent this overloading.

A short circuit is a direct contact or connection between a live conductor and (*a*) a neutral or return conductor or (*b*) earthed metalwork, the contact usually being the result of an accident. The result of the short-circuit is to present a conducting path of extremely low resistance which will allow the passage of a current often of many hundreds of amperes. If the faulty circuit has no overcurrent protection, the cables will heat up rapidly and melt, equipment would also suffer severe damage and fire would be the inevitable result.

Apart from the relays associated with circuit-breakers mentioned already, two methods of overcurrent protection are in wide use: fuses and circuit-breakers. The latter, so far as domestic and small industrial loads are concerned, are miniature circuit-breakers (MCBs).

Fuses

A fuse is defined as 'A device for opening a

circuit by means of a conductor designed to melt when an excessive current flows along it. The fuse comprises all the parts of the complete device.' Other terms relating to the fuse are:

(*a*) *Fuse-element*. That part of a fuse which is designed to melt and thus open a circuit.

(*b*) *Cartridge fuse*. A fuse in which the fuse-element is totally enclosed in a cartridge.

(*c*) *Fuse-link*. That part of a fuse which comprises a fuse-element and a cartridge or other container, if any, and either is capable of being attached to fuse-contacts or is fitted with fuse-contacts as an integral part of it.

There are three main types of fuse: the rewirable, the cartridge and the HBC (high breaking capacity) fuse; the latter is a development of the cartridge type.

The rewirable type of fuse consists of a porcelain (usual material) bridge and base. The bridge has two sets of contacts which fit into other contacts in the base. The fuse-element, usually tinned copper wire, is connected between the terminals of the bridge. An asbestos tube or pad is usually fitted to reduce the effects of arcing when the fuse-element melts.

Three terms are used in connection with fuses:

Current rating. This is the maximum current that a fuse will carry indefinitely without undue deterioration of the fuse-element.

Fusing current. This is the minimum current that will 'blow' the fuse.

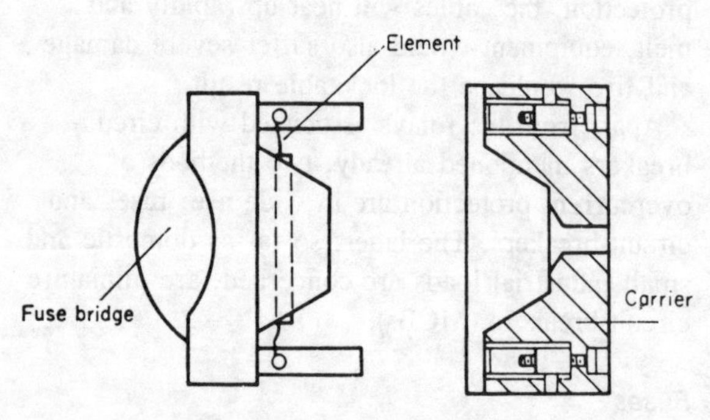

Figure 15.2. Typical rewirable fuse.

Fusing factor. This is the ratio of the minimum fusing current to the current rating, namely

$$\text{Fusing factor} = \frac{\text{minimum fusing current}}{\text{current rating}}$$

The rewirable fuse (BS 3036) is a simple and relatively cheap type of over-current protective device and is still widely used despite several disadvantages including:

(*a*) The ease with which an inexperienced person can replace a 'blown' fuse-element with a wire of incorrect gauge or type.

(*b*) Undue deterioration of the fuse-elements due to oxidisation.

(*c*) Lack of discrimination. This means that it is possible, in certain installation conditions, for a 15 A fuse-element to melt before a 10 A fuse-element. Also, a rewirable fuse is not capable of discriminating between a momentary high current (e.g. motor starting current) and a continuous fault current.

(*d*) Damage, particularly in conditions of severe short-circuit.

The fusing factor for a rewirable fuse is about 2. With a protective asbestos pad it is about 1.9. This means that a fuse-element rated at 10 A will melt when $10 \times 2 = 20$ A flows in the circuit. This also means that, say, a 1.00 mm^2 conductor (which has a current rating of 14 A) may, in an overload condition, be made to carry as much as 50 per cent overload without the fuse coming into action; the cable is thus run on overload which may lead eventually to damage to its insulation.

The obvious disadvantages of the rewirable type of fuse led to the development and use of the cartridge fuse which is most often found in 13 A fused plugs. Figure 15.3 shows the construction of

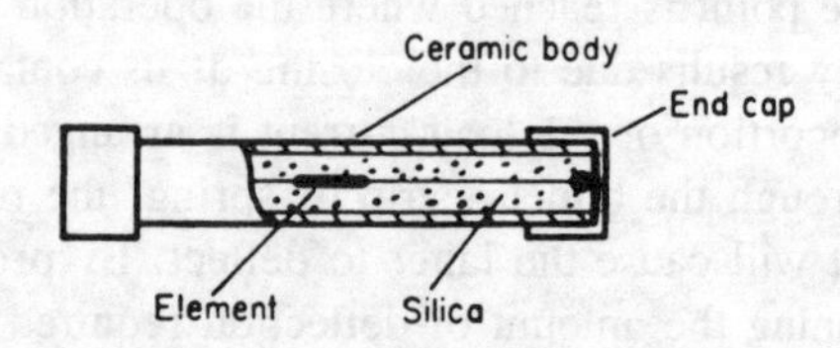

Figure 15.3. Typical cartridge fuse.

the cartridge fuse. The fusing factor of this type is about 1.5. Thus, a cartridge fuse rated at 13 A will 'blow' at 19 A. The rating of this type of fuse is determined and fixed by the manufacturer. Among the advantages of the cartridge fuse are: (*a*) the current rating is accurately known and (*b*) the fuse-element is less liable to deterioration in service. Two main disadvantages are: (*a*) the fuse-element is more expensive to replace than the rewirable type and (*b*) it is unsuitable for use where extremely high values of fault current may occur.

So far as small domestic and industrial loads are concerned, the following types or ratings are available:

1. House-service cutout fuselinks (to BS 88) for use in supply authority cutouts instead of rewirable fuses.
2. Ferrule-cap fuselinks (to BS 1361) for use in domestic 250 V consumer control units, switchfuses, switch splitters, etc. The ratings and identification colours are:

Rating (amps)	*Colour code*
5	white
15	blue
20	yellow
30	red
45	green
60	purple

3a. Domestic cartridge fuselinks (to BS 1363) specifically for use with 13 A fused rectangular-pin plugs. The ratings are 3 A and 13 A.

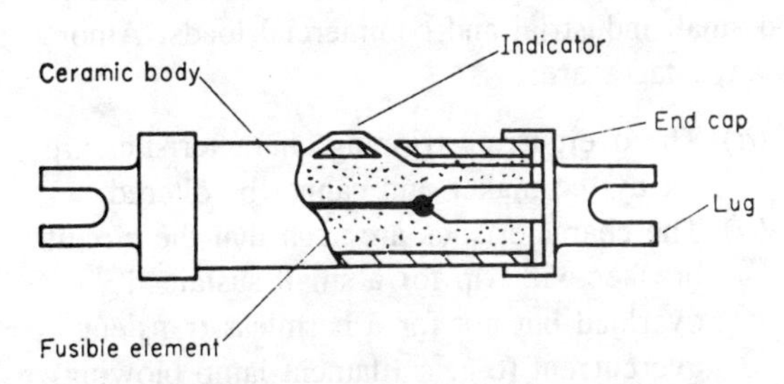

Figure 15.4. Typical HRC cartridge fuse.

3b. Domestic cartridge fuselinks (to BS 646) specifically for use with 15 A, round-pin plugs, where the load taken from the 15 A socket-outlet is small (e.g. radio, or a table lamp) in relation to the 15 A fuse protecting the subcircuit. The ratings and identification colours are:

Rating (amps)	*Colour code*
1	green
2	yellow
3	black
5	red

Note that all these cartridge fuses are so designed that they are not interchangeable except within their own group.

The high breaking capacity fuse (HBC) has its fusing characteristic carefully controlled by the manufacturer. As its name suggests it can safely interrupt very large fault currents. The fuses are often used to protect large industrial loads and main cables. Figure 15.4 shows the construction of a typical HBC fuse. The cartridge barrel is of high-grade ceramic able to withstand the shock conditions when a heavy fault current is interrupted. The end caps and fixing tags are suitably plated to give good electrical contact. The fixing tags are planished when necessary to ensure satisfactory alignment between contact-mating surfaces. Except for very low ratings, the fuse-element is made from pure silver. It has an accurately machined waist to ensure consistency and reliability. The shape of the waist is designed to give the required operational characteristic.

The filler is powdered silica, carefully dried before use. When used, the filler is compacted in the barrels by mechanical vibration to ensure complete filling. An indicator is provided to show when the fuse has blown. It consists of a glass bead held in position in a recess in the barrel by a fine resistance wire, connected in parallel with the fuse-elements. Barrels are accurately ground and the caps are a force fit. Correct grades of solder are used for the element and tag fixings. The larger types of multi-element fuses have the elements welded in addition to soldering.

The HBC fuse is more expensive than either the rewirable type or the cartridge type. The fusing factor of an HBC fuse is, for small loads, up to 1.25. Thus, a 10 A HBC fuse will blow at 10 × 1.25 = 12.5 A. HBC fuses are discriminating, which means that they are able to distinguish between a high starting current taken by a motor (which lasts only a matter of seconds) and a high fault or overload current (which lasts longer). HBC fuses are often used in motor circuits for 'back-up' protection for the machines. Motors are normally protected against overload by the starter trip; the fuses are required only to give protection against short-circuit currents and overloads outside the capacity of the thermal trip. Modern squirrel-cage induction motors can take up to eight times normal full-load current when stalled. The rating of a fuse-link for a motor circuit should be the smallest current rating that will carry the starting current while providing the necessary margin of safety.

When a capacitor is switched into a circuit, a heavy inrush of current will flow. To ensure that fuses do not blow unnecessarily in these circumstances, it is necessary to fit higher rated fuses. In general, if the fuses fitted are rated at 125–150 per cent of the capacitor rating, nuisance blowing of the fuses will be avoided. Transformer and fluorescent lighting circuits may also need higher rated fuse-links to deal with the inrush currents associated with this class of gear. Fuselinks with a rating of about 50 per cent greater than the normal current of the apparatus to be protected are usually found to be satisfactory.

One point to remember about fuses and fuse protection is that circuit fuses protect the circuit cables from being overloaded and should also prevent main fuses operating in the case of a local short-circuit. Circuit fuses do not protect a current-using device from becoming overloaded, especially when a circuit has more than one outlet.

Circuit-breakers

The circuit-breaker is described in Chapter 20 'Circuit-control devices'. Briefly, it is an automatic device for making and breaking a circuit both under normal and abnormal conditions, such as those of a short circuit. The circuit-breaker has several advantages over any type of fuse:

(*a*) In the event of a fault or overload, all poles are simultaneously disconnected from the supply.
(*b*) It is capable of remote control by means of emergency stop-buttons.
(*c*) Overload and time-lags are capable of adjustment within limits.
(*d*) The circuit can be closed again quickly onto the fault safely.
(*e*) It can open a circuit if the supply fails, thus avoiding unexpected reintroduction of the supply causing apparatus to become live.

It is generally required that every circuit-breaker shall open a circuit before the current in the circuit exceeds twice that of the rating of the smallest cable it protects (this does not apply to motor circuits). For motor circuits, a time-lag is arranged so that the heavy starting current can be carried for a short period — and satisfactory protection is provided for in normal running conditions.

The usual circuit-breaker arrangement is one in which an overcurrent is used to open the switch. The switch mechanism is fitted with an arrangement of springs which are compressed as the contacts are closed. Once closed, the mechanism is held in position against the pressure of the springs by a gravity-operated trigger. When released, this trigger causes the mechanism to collapse and open the switch. Solenoid coils are used to carry the maximum current, or some proportion of it, flowing in the controlled circuit. When this current is exceeded the solenoid plunger moves to trip the circuit-breaker.

Recent years have seen the rapid development of the miniature circuit-breaker (MCB) as an alternative to fuses as a means of protection for domestic and small industrial and commercial loads. Among its advantages are:

(*a*) The overcurrent tripping characteristics are set by the maker and cannot be altered.
(*b*) The characteristics are such that the circuit-breaker will trip for a small sustained overload but not for a harmless transient overcurrent (e.g. a filament lamp blowing). Operation is instantaneous when a short-circuit current flows.

(*c*) Faulty circuits are easily identified with the ON and OFF position of the device.
(*d*) The supply can be quickly and easily restored when the fault has been cleared. And if the MCB is switched while the fault is present, it will still be able to 'clear' itself and open satisfactorily.

A larger version of the MCB is the moulded-case circuit-breaker (MCCB) which is used for main control and protection and is available for much higher ratings (e.g. 400 A) than are the MCBs which are used for final-circuit protection and have ratings up to 63 A. It is important to install MCBs with an appropriate short-circuit rating. Typical ratings are 6 kA, 9 kA and 16 kA. It should also be noted that there are three types of MCB, each with its own operating times.

16 Earthing

The purpose of earthing is to ensure that no person operating an electrical installation can receive an electric shock which could cause injury or a fatality. In simple terms, 'earthing' involves the connection of all metalwork associated with the electrical installation (the exposed conductive parts) with protective conductors (CPCs) which are terminated at a common point, the main earth terminal. This terminal is further connected to a proven earth connection which can be the supply authority's wire-armoured supply cable, an overhead line conductor or an earth electrode driven directly into the soil. The availability of one or other of these connections depends on the type of electrical system used to supply electricity.

Apart from the 'exposed conductive parts' found in an installation, there is other metalwork which has nothing to do with the electrical installation but which could become live in the event of a fault to earth. This metalwork is known as 'extraneous conductive parts' and includes hot and cold water pipes, radiators, structural steelwork, metal-topped sink units and metallic ducting used for ventilation. These parts are connected by means of (*a*) main bonding conductors and (*b*) supplementary bonding conductors. The former are used to bond together metallic services at their point of entry into a building. The latter are used to bond together metallic pipes and the like within the installation. These bonding conductors are also taken to the installation's main earth terminal. Thus all metalwork in a building is at earth potential.

Once all CPCs and bonding conductors are taken to the main earth terminal, the building is known as an 'equipotential zone' and acts as a kind of safety cage in which persons can be reasonably assured of being safe from serious electric shock. Any electrical equipment taken outside the equipotential zone, such as an electric lawnmower, must be fed from a socket-outlet which incorporates a residual current device (RCD). The word 'equipotential' simply means that every single piece of metal in the building is at earth potential.

The earthing of all metalwork does not complete the protection against electric shock offered to the consumer. Overcurrent devices (fuses and MCBs) are required to operate within either 0.5 second or 4 seconds if a fault to earth occurs. And the use of RCDs also offers further protection in situations when an earth fault may not produce sufficient current to operate overcurrent protective devices.

Even before the days of electricity supply on a commercial scale, the soil has been used as a conductor for electrical currents. In early telegraphy systems the earth was used as a return conductor. The early scientists discovered that charges of electricity could be dissipated by connecting a charged body to the general mass of earth by using suitable electrodes, of which the earliest form was a metal plate (the earth plate). But the earth has many failings as a conductor. This is because the resistance of soils varies with their composition. When completely dry, most soils and rocks are non-conductors of electricity. The exceptions to this are, of course, where metallic minerals are present to form conducting paths. Sands, loams and rocks can therefore be regarded as non-conductors; but when water or moisture is present, their resistivity drops to such a low value that they become conductors — though very poor ones. This means that the resistivity of a soil is determined by the quantity of water present in it — and on the

Description	ohm metre
Marshy ground	2 to 3.5
Loam and clay	4 to 150
Chalk	60 to 400
Sand	90 to 8,000
Peat	50 to 500
Sandy gravel	50 to 500
Rock	1,000 upwards

Figure 16.1. Resistivity of soils.

resistivity of the water itself. It also means that conduction through the soil is in effect conduction through the water, and so is of an electrolytic nature. Figure 16.1 shows some typical values of resistivity for some soils.

For all that the earth is an inefficient conductor, it is widely used in electrical work. There are three main functions of earthing:

1. To maintain the potential of any part of a system at a definite value with respect to earth.
2. To allow current to flow to earth in the event of a fault, so that the protective gear will operate to isolate the faulty circuit.
3. To make sure that, in the event of a fault, apparatus normally 'dead' cannot reach a dangerous potential with respect to earth (earth is normally taken as 0 V, 'no volts').

IEE Regulation 130–04 states that where metalwork, other than current-carrying conductors, is liable to become charged with electricity in such a manner as to create a danger if the insulation of a conductor should become defective, or if a defect should occur in any apparatus (i) the metalwork shall be earthed in such a manner as will ensure immediate electrical discharge without danger. . . . Chapter 41 of the IEE Regulations details how the above can be complied with. The basic reason for earthing is to prevent or to minimise the risk of shock to human beings. If an earth fault occurs in an installation it means that a live conductor has come into contact with metalwork to cause the metalwork to become live — that is, to reach the same potential or voltage as the live conductor. Any person touching the metalwork, and who is standing on a non-insulating floor, will receive an electric shock as the result of the current flowing through the body to earth. If, however, the metalwork is connected to the general mass of earth through a low-resistance path, the circuit now becomes a parallel-branch circuit with:

(*a*) the human body as one branch with a resistance of, say, 10 000 ohms; and
(*b*) the CPC fault path as the other branch with a resistance of 1 ohm or less.

The result of properly earthed metalwork is that by far the greater proportion of fault-current will flow through the low-resistance path, so limiting the amount of current flowing through the human body. If the current is really heavy (as in a direct short circuit) then a fuse will blow or a protective device will operate. However, an earth fault-current may flow with a value not sufficient to blow a fuse yet more than enough to cause over-heating at, say, a loose connection to start a fire.

Regulations

Chapter 41 of the IEE Regulations deals with the requirements which all earthing arrangements must satisfy if an electrical installation is to be deemed safe. The main basic requirements are:

1. The complete insulation of all parts of an electrical system. This involves the use of apparatus of 'all-insulated' construction, which means that the insulation which encloses the apparatus is durable and substantially continuous.
2. The use of appliances with double insulation conforming to the British Standard Specifications.
3. The earthing of exposed metal parts (there are some exemptions).
4. The isolation of metalwork in such a manner that it is not liable to come into contact with any live parts or with earthed metalwork.

The basic requirements for good earthing are that the earthing arrangements of the consumer's installation are such that the occurrence of a fault of negligible impedance from a phase or non-earthed conductor to adjacent exposed metal, a current corresponding to three times the fuse rating or 1.5 times the setting of an overcurrent circuit-breaker can flow, so that the faulty circuit is made dead. The earthing arrangement should be such that the maximum sustained voltage developed under fault conditions between exposed metal required to be earthed and the consumer's earth terminal should not exceed 50 V.

The IEE Regulations detail the metalwork found in premises, called 'extraneous conductive parts', which are required to be connected to the installations main earthing terminal. Bathrooms and showers are also covered.

The CPC is the conductor which bonds all metalwork required to be earthed. If it is a separate conductor (insulated and coloured green) it must be at least 1/1.13 (csa = 1.00 mm^2) and need not be greater than 70 mm^2. Note that conduit and trunking may be used as the sole CPC except in agricultural installations.

Where metal conduit is used as a CPC, a high standard of workmanship in installation is essential. Joints must be really sound. Slackness in the joints may result in deterioration in, and even complete loss of, continuity. For outdoor installations and where otherwise subjected to atmospheric corrosion, screwed conduit should always be used, suitably protected against such corrosion. In screwed conduit installations, the liberal use of locknuts is recommended. Joints in all conduit systems should be painted overall after assembly. In mixed installations (e.g. aluminium-alloy conduit with steel fittings, or steel conduit with aluminium-alloy or zinc-base-alloy fittings) the following are sound recommendations to ensure the electrical continuity of joints.

All threads in aluminium or zinc alloys should be cut using a suitable lubricant. A protective material (e.g. petroleum jelly) should be applied to the threads in all materials when the joints are made up. All joints should be made tight. The use of locknuts is advised. In addition, it is recommended to apply bituminised paint to the outside of all joints after assembly. In damp conditions, electrolytic corrosion is liable to occur at contacts between dissimilar metals. To avoid this, all earthing clamps and fittings in contact with aluminium-base-alloy tubing should be of an alloy or finish which is known from experience to be suitable. Copper, or alloys with a high copper content, are particularly liable to cause corrosion when in contact with aluminium-base alloys. For this reason, brass fixing screws or saddles should not be used with conduit or fittings of aluminium-base alloys. Periodical tests should be made to ensure that electrical continuity is satisfactorily maintained. Flexible metal conduits should not be used as a CPC. Where flexible tubing forms part of an earthed metal conduit system, a separate copper or copper-alloy CPC should be installed with the tubing and connected to it at each end.

The earth-conductor lead should be of a minimum size: 6 mm^2, except 2.5 mm^2 is accepted for connection to an earth-leakage circuit-breaker. It must also be protected against mechanical damage and corrosion, and not less than half the largest size of the conductor to be protected, but need not normally exceed 70 mm^2.

There are a number of methods used to achieve the earthing of an installation:

1. Connection to the metal sheath and armouring of a supply authority's underground supply cable.
2. Connection to the continuous earth wire (CEW) provided by a supply authority where the distribution of energy is by overhead lines.
3. Connection to an earth electrode sunk in the ground for the purpose.
4. Installation of a protective-multiple earthing system.
5. Installation of automatic fault protection.

One disadvantage in using a mains water-pipe is that sections of the pipe may be replaced by sections of non-conducting material (PVC or asbestos), which makes the pipe an inconsistent earth electrode. The provision of a cable sheath as an earthing connection (method (2)) is very common nowadays. Usually, however, it is accepted that if, for any reason, the earthing is subsequently proved ineffective, the supply authority is not to be made responsible. Continuous earth wires are not always provided by the supply authority, except in those areas which have extremely high values of soil resistivity (e.g. peat and rock). The CEW is sometimes called an aerial earth. Connection to an earth electrode sunk in the ground is the most common means of earthing. The earth electrode can be any one of the following forms:

(a) Pipe. Generally a 200 mm diameter cast-iron pipe, 2 m long and buried in a coke-filled pot. This type requires a certain amount of excavation; iron is, of course, prone to corrosion, particularly if the coke has a high sulphur content.

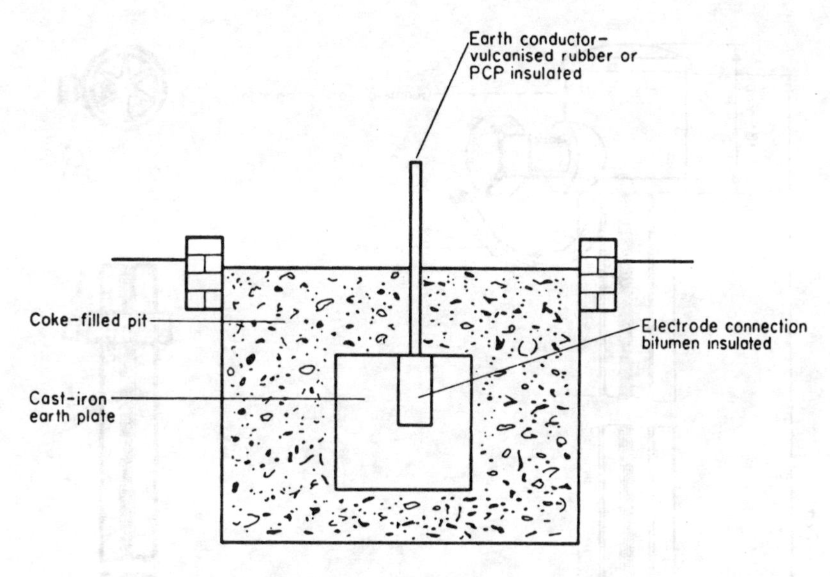

Figure 16.2. Typical earth-electrode pit.

(b) Plate. Plate electrodes are normally of cast-iron, buried vertically with the centre about 1 m below the surface. Copper plates may also be used. Plate electrodes provide a large surface area and are used mainly where the ground is shallow (where the resistivity is low near the surface but increases rapidly with depth). Again, excavation is required. Care is needed to protect the earth-electrode connection (to the earthing lead) from corrosion.

(c) Strip. Copper strip is most useful in shallow soil overlying rock. The strip should be buried to a depth of not less than 50 cm.

(d) Rods. Rod electrodes are very economical and require no excavation for their installation. Because buried length is more important than diameter, the extensible, small-diameter copper rod has many advantages. It can, for instance, be driven into the ground so that the soil contact with the rod is close and definite. Extensible rods are of standard lengths and made from hard-drawn copper. They have a hardened steel tip and a steel driving cap. Sometimes the copper rod has a steel rod running through its centre for strength while it is being driven into rocky soil. Ribbed earth rods have wide vertical ribs to give a high degree of mechanical stiffness, so that they are not easily bent or deflected when driven into the ground.

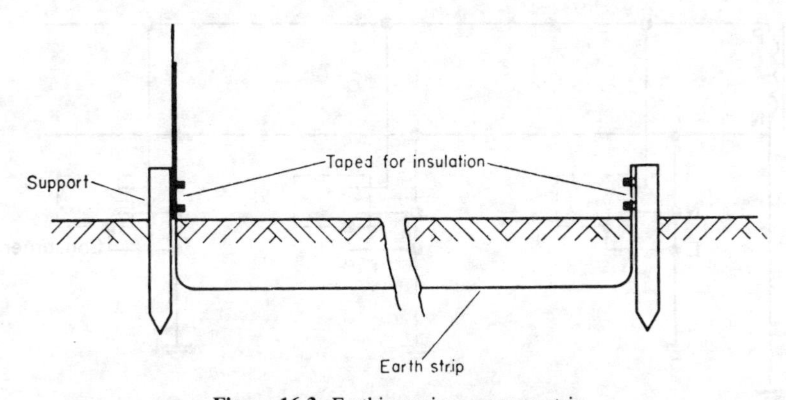

Figure 16.3. Earthing using a copper strip.

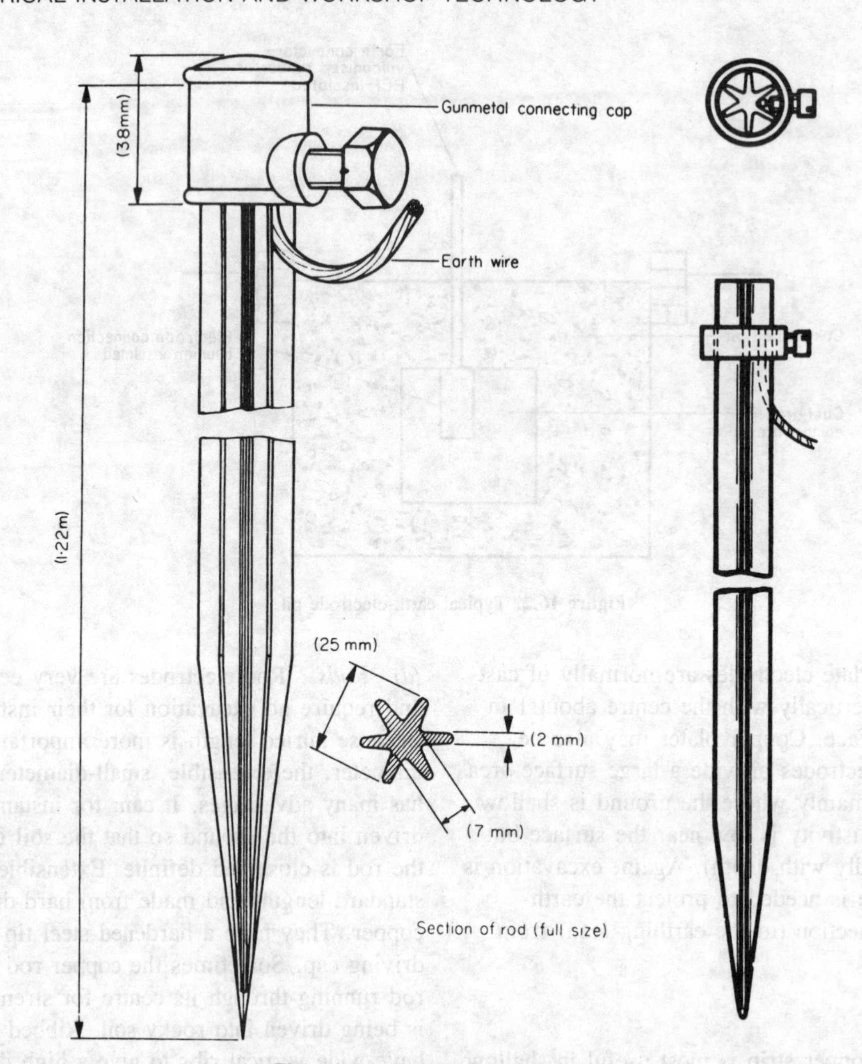

Figure 16.4. Ribbed earth-electrode.

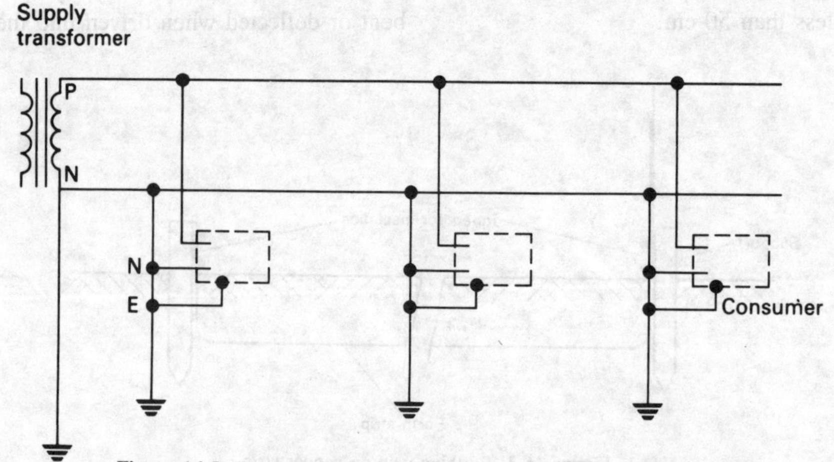

Figure 16.5. Diagram showing the Protective Multiple Earthing system.

Because the method used to connect the earthing lead to the earth electrode is important, all clamps and clips must conform to the requirements of the IEE Regulations.

The PME method (Regulation Appendix 2) gives protection against earth-fault conditions and uses the neutral of the incoming supply as the earth point or terminal. In this system of earthing, all protected metalwork is connected, by means of the installation CPCs, to the neutral-service conductor at the supply-intake position. By doing this, line-to-earth faults are converted into line-to-neutral faults. The reason for this is to ensure that sufficient current will flow under fault conditions to blow a fuse or trip an overload circuit-breaker, so isolating the faulty circuit from the supply.

The PME system has a number of disadvantages and stringent requirements are laid down to cover the use of the system (see Appendix 2 in the IEE Regulations). Figure 16.5 shows a typical distribution system with consumers connected to a common PME system of earthing.

Residual-current ELCBs are now only recognised by the Regulations. The basic principle of operation depends upon more current flowing into the live side of the primary winding than leaves by the neutral, or other return (earth) conductor. The essential part of the residual-current ELCB is a transformer with opposed windings carrying the incoming and outgoing current. In a healthy circuit, where the values of current in the windings are equal, the magnetic effects cancel each other out. However, a fault will cause an out-of-balance condition and create a magnetic effect in the transformer core which links with the turns of a small secondary winding. An emf (electromotive force measured in volts) is induced in this winding. The secondary winding is permanently connected to the trip coil of the circuit-breaker. The induced emf will cause a current to flow in the trip coil: if this current is of sufficient value the coil will become energised to trip the breaker contacts. A test switch is provided.

Earth-fault-loop path

Figure 16.7 shows the path taken by an earth-fault current. The resistance symbols in both the consumer's earth electrode and the neutral earth electrode indicate the equivalent resistance of the earthing connections.

It should be noted in Figure 16.7 that the consumer's earth electrode can also be the armouring of the supply authority's underground supply cable.

As will be appreciated, all the parts of the earth-fault loop path have resistance, more usually called 'impedance'. Both terms indicate that the flow of

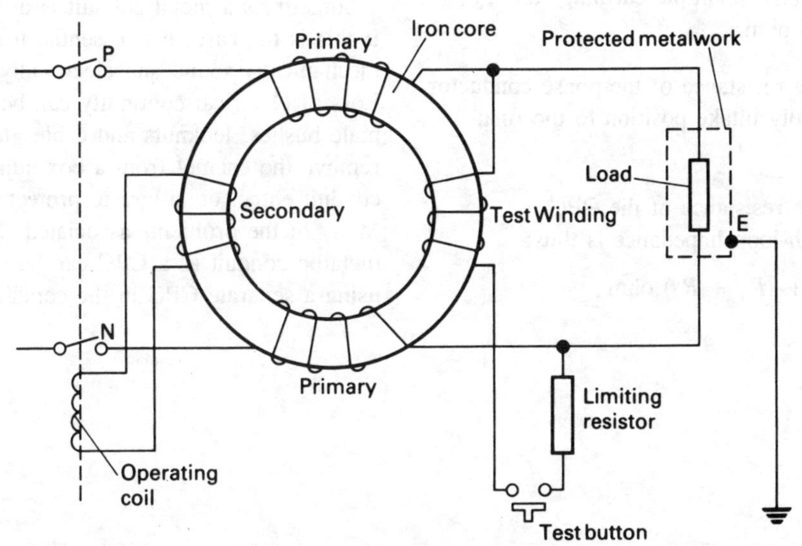

Figure 16.6. Circuit diagram of a residual-current earth-leakage circuit-breaker.

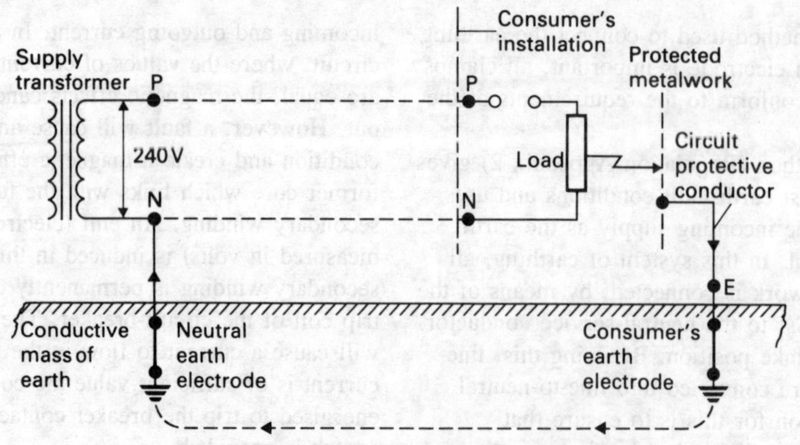

Figure 16.7. Typical earthing arrangement of a single-phase supply.

current is resisted or impeded. The unit used in both cases is the same: the ohm. Thus the total sum of the impedance of the individual parts of the path is known as the loop impedance, Z_s. This is made up of:

Z_e, which is the impedance external to the consumer's intake terminals and includes the supply transformer secondary winding, the phase conductor and the metallic return path back to the supply transformer earthing arrangement. The value of Z_e depends on the type of supply system and is available from the supply authority. For example, if the supply system is TNCS (protective multiple earthing) the Z_e value is 0.35 ohm.

R_1, which is the resistance of the phase conductor from the supply intake position to the final circuit load.

R_2, which is the resistance of the CPC.

The total earth-loop impedance is thus:

$$Z_s = Z_e + (R_1 + R_2) \text{ ohm}$$

The maximum value for Z_s depends on the type of overcurrent protective device and its rating and is also dependent on whether the final circuit supplies socket-outlets or fixed equipment. In any case when a fault to earth occurs, the circuit should be disconnected within either 0.4 second or 5 seconds. For instance, a final circuit feeding socket-outlets and protected by a 20 A BS 1361 fuse will have a maximum value for Z_s of 1.8 ohms. In the event of an earth fault, the current which would flow would be $I_f = V/Z_s$. Assuming a 240 V supply, the value of the current would be 133 A, which is enough to ensure the fuse would blow within the 0.4 second limit.

Sometimes a metal conduit is used as the CPC. If this is the case, it is essential that all joints be mechanically sound and electrically continuous. Loss of electrical continuity can be traced to slack male bushes, locknuts and cable glands, failure to remove the enamel from a box adjacent to a conduit entry, or failure to protect against rust. Many of the problems associated with the use of metallic conduit as a CPC can be eliminated by using a separate CPC in the conduit.

17 Testing and measuring instruments

Whatever the quality of materials used in an installation and the quality of workmanship used to install the materials, accessories and equipment which all go to make up an electrical installation, the complete job must be both inspected and tested. The details of this requirement of the Wiring Regulations are outlined in Chapter 18. This chapter describes the types of instruments used to carry out the various tests. A look is also taken at some other types of instruments with which the practising electrician is expected to be familiar which are used to measure electrical quantities rather than perform test functions.

Recent years have seen a significant advance in the development of instruments used for testing purposes. Many use electronic devices and circuitry and are designed to be easily operated and easily read. Some instruments use the familiar graduated scale across which a needle pointer moves to indicate a reading; others use a digital read-out display and are often more accurate.

All instruments have to be checked before they are used, particularly where installation tests are carried out. Battery-operated testers should be subjected to a pre-test procedure to ensure that the instrument is able to give a reliable reading.

The batteries must be checked to ensure they are in a good condition. Most testers incorporate a 'good' or 'poor' indicator. If the batteries are healthy, the instrument should be checked for accurate readings against a known-value resistor where this is appropriate. For example, an insulation-resistance tester should be checked against a 1 megohm resistor. If these two conditions are satisfied, healthy batteries and accuracy of reading, the instrument can then be used on the test.

All test instruments should be recalibrated after a period of time. And tests should be made with the same type (or model) of instrument to ensure that readings obtained at subsequent tests accord with each other. It is often required that the serial number of the test instrument is recorded so that the same tester can be used at a later time for a similar test.

Insulation-resistance tester

There are two types: battery-operated and those incorporating a small hand-operated generator. These instruments produce a voltage which is twice the normal working voltage of the circuit under test, e.g. 500 V for a 240 V system. Other test voltages include 110 V and 1000 V. Most insulation-resistance testers have two scales: megohm and ohm. The former indicates the value of insulation resistance between conductors and between conductors and earth, with the scale opened out, as it were, around the minimum acceptable value for circuits: 1 megohm. The other scale, reading in ohms, is used for checking the conductor continuity of circuits, with the beginning of the scale again opened out for accurate readings. Digital-display instruments give the actual value of a test reading.

The reason for the test voltage being twice the working voltage is to stress the insulation. If it has been damaged or has deteriorated, the test will show this condition. The test current actually flows through the insulation material and so any faults will readily be detected: the better the insulation the less current will flow.

Continuity tester

This instrument indicates low ohmic values, typically 0–100 ohms. It is often used to measure the resistance of conductors where accuracy is not required. If readings are expected much less than 1 ohm, a milliohmmeter is used which, typically, can measure resistance down to 1 milliohm. Some installation tests (e.g. ring-circuit continuity and

CPC continuity) are best measured using a milliohmmeter. Most are battery operated and some instruments have a digital display.

Phase-earth loop tester

This instrument is used to check the value of the impedance of the earth-loop path (Z_s). It passes, through a current-limiting resistor, a current of about 25 A into the earth circuit and the reading obtained, in ohms, is then compared with a maximum value of Z_s indicated in tables (Chapter 41) in the Wiring Regulations. The maximum value depends on the type of protective device used for the circuit under test, and its current rating. All testers have two indicating lights which must be ON before the test is carried out. The L–N light indicates that the polarity of the circuit is correct. The L–E light verifies that there is a connection between phase and earth. If one or both of these lights do not operate, the test must not be carried out. Testers are available with a digital read-out or with a graduated scale in ohms.

RCD tester

This instrument checks the operating (tripping) current of RCDs. Most instruments offer a range of tripping currents (typically 5 mA–500 mA) and some indicate how fast the RCD has tripped (i.e. time of disconnection).

Earth tester

This instrument is used to check the effectiveness of the resistance to earth of each element of an earth installation and, in particular, where an independent earth electrode is used as the main means of earthing (e.g. in a TT earthing arrrangement). The test can also be used to measure earth continuity.

Portable appliance tester

This instrument is used to check the electrical condition of portable appliances. Three test facilities are incorporated: earth bond, insulation resistance (IR) and a flash test. The first test facility ensures that the earthing provision (e.g. CPC in the flex and the metal casing of the appliance) is adequate enough to carry a severe earth-fault current. The IR test checks the condition of the insulation in both the flexible cord and the appliance. The flash test is used for double-insulated appliances, which is stressed at a voltage up to 5 kV to verify the integrity of the insulation used in the construction of the appliance.

Measuring instruments

These are used to measure voltage, current and resistance. When checking the voltage of a live circuit, it is important to ensure one's own safety by using fused safety probes to make contact with

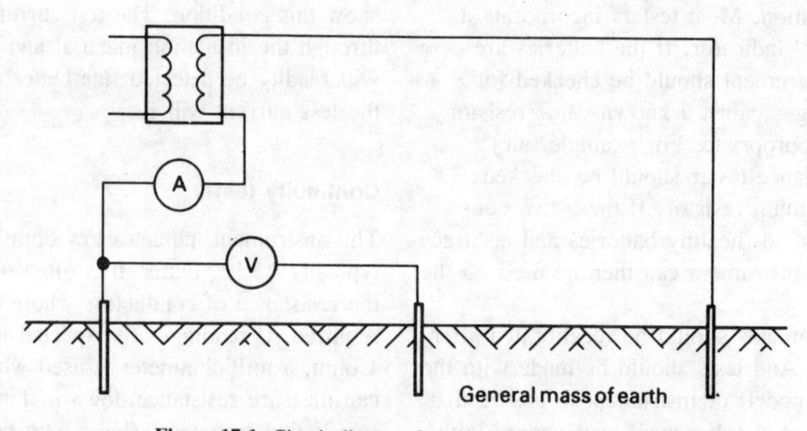

Figure 17.1. Circuit diagram for testing soil resistivity.

the live terminals. Instead of using a voltmeter, a potential (voltage) indicator can be used to check the voltage range of a circuit. These are safer to use.

Ammeters are normally connected into a circuit, which also raises the question of safety. On ac circuits, the clamp meter does away with the need to open up a circuit conductor to measure the level of current flow. These instruments have insulated jaws which are clamped round the insulated conductor; current values are then read off a digital read-out or off a graduated scale. Most clamp meters offer a wide range of current values and some incorporate a voltage reading facility.

Resistance can be measured with an ohmmeter which is battery operated. Some instruments are available which measure current, voltage and resistance on both ac and dc supplies. When using these multi-range instruments care must be taken to select the correct electrical quantity to be measured and the correct scale.

Recording of the test results

Any test result obtained from a test on an installation must be recorded as it then becomes part of the history of the installation. If a similar test has to be carried out at a later date, the person carrying out the new test can then compare the new value with the recorded value. For instance, if an earth-loop test revealed an impedance (Z_s) value of 1.8 ohms which was an acceptable reading and a later test indicated, say, 2 ohms, which was still within the maximum Z_s limit, the reason for the increase should be investigated as the second reading might indicate a deterioration in part of the earth-loop path. A more than likely culprit might be the CPC connection/termination which might have become insecure. If the CPC was metal conduit, the increase might be due to slack locknuts or rusting at the conduit joints.

Except where circuits and electrical equipment are tested 'dead', all precautions must be taken to ensure safety from electric shock and burns. Attention is drawn to the Health and Safety Executive Guidance Note G538 *Electrical Test Equipment for Use by Electricians*. This publication advises on the selection and safe use of suitable test probes, leads, lamps, voltage-indicating devices and other measuring equipment.

Live electrical testing is generally limited by the Electricity (Factories Act) Special Regulations to 'authorised' or 'competent' persons where the voltage involved in more than 125 V ac (or 250 V dc); and these persons must be over the age of twenty-one years and under the immediate supervision of the authorised person. This age limit for a competent person may be reduced to eighteen years (under the Electricity Regulations (Competent Persons) Order 1968) provided that he/she has completed an approved training course.

18 Inspection and testing

The Wiring Regulations require that every new installation, an extension or alteration to an existing installation, must be inspected and tested on completion and before it is energised. This is to give the client some guarantee that 'good workmanship and the use of approved materials' will provide an installation which is safe to use and which, in all respects, reduces to a minimum all risks against electric shock, fire and burns.

Before any inspection and test can be carried out, certain information must be available to the inspector. This information includes drawings showing the positions of isolators, switches, outlet points, the type and rating of protective devices and the size and type of conductors. In other words, much of the information which relates to the design of the installation has to be made available. It is generally accepted in the trade that not only is the designer responsible for the quality built into the installation, but the installer (the electrician) bears some responsibility for how the installation is presented. In addition, the person carrying out the inspection and test must be fully familiar not only with the instruments used in testing but also be fully competent in their use — and be able to interpret the test results.

Visual inspection

The visual inspection of a complete installation includes checking the connections of conductors and their correct identification (by colour or numbers). The conductors must also be checked to ensure that they are able to carry the designed load current and that the voltage drop is within specified limits. All single-pole control devices (switches) and protective devices (fuses, MCBs) must be connected to the phase conductors. 'Polarised' wiring accessories (such as socket-outlets and Edison screw (ES) lampholders) must be correctly terminated.

In work areas, there must be provided switches for isolation (e.g. in a motor circuit) and emergency switching. All distribution boards and consumer units must have circuit charts giving details of the function of final circuits, their conductor sizes and the type and rating of the protective devices. In situations where there are a number of switches, distribution boards, etc. these must be identified as to their purpose with labels. Danger and warning notices must also be provided (e.g. where high voltages are used for external lighting installations and for firemens' switches).

The visual inspection must also include checks to make sure that there is no danger to those using the installation from direct contact with live terminals.

The Wiring Regulations publish a check-list which is not comprehensive but gives the inspector a good starting-point from which to draw up his/her own check-list which, of course, will depend on the type of installation. The inspector must also check such simple things like: no burrs are left inside conduit; all cables are protected against mechanical damage; cables are securely fixed with correct distances between their supports; and all materials used in the installation have been correctly chosen to operate in the working conditions which obtain once the installation has been energised.

Testing

The range of testing instruments available is described in the previous chapter. It is, however, worth mentioning that the instruments must be in good condition, be properly calibrated and used correctly. The tests which are required by the Wiring Regulations to be carried out do not apply to every installation. For instance, the test schedule for a domestic installation will differ from that for an industrial installation. The following list is confined to the tests applied to a domestic installation:

1. Continuity of socket-outlet final ring circuit conductors;
2. Continuity of CPCs, including main and supplementary bonding contractors;
3. Earth electrode resistance;
4. Insulation resistance;
5. Polarity;
6. Earth-fault loop impedance;
7. Operation of RCDs.

All test results must be logged so that they can be used for comparison with the results obtained at future tests. This is done to provide a historical record of the condition of the installation as it grows older and so that potential dangers can be rectified before increased risks from electric shock and fire can reach a dangerous level.

Continuity of ring final circuit conductors

This test is made to ensure the electrical continuity of the conductors of ring circuits. Ideally, there should be no break in these conductors. Normal practice, however, means that conductors are 'broken' and terminated in terminals and secured by screws. These in fact constitute what are called 'dry joints' which, through time, could become loose. This is easily demonstrated with the terminals of a 13 A plug which is used to supply a 3 kW washing machine. Over a period of time the current taken by the appliance will tend to produce heat which results in the continual expansion and contraction of the terminal screws which eventually become loose. After a while it is possible to get an extra half-turn or so on the screw to retighten it. If this happens in a 13 A plug, it can also occur at the terminals behind the socket-outlet.

The test can be carried out using the 'continuity' range of a dual-purpose instrument (e.g. an insulation-resistance tester) or using a milliohm-meter. The latter is preferred because the instrument can give readings as low as 1 milliohm.

The test is made to ensure that the conductors (phase, neutral and the CPC) are electrically continuous and to make sure that there has been no inadvertent interconnection between socket-outlets which would constitute a short-circuit which in effect creates an apparently continuous ring circuit. In this situation an actual break in a conductor could exist without detection.

The test is carried out as follows:

A Measure the resistance of the conductor (e.g. phase) by separating its ends at the consumer unit or distribution board. Call the reading obtained *A*.

B Join the separated ends of the conductor and connect the instrument to the join. The other terminal of the instrument is then connected to the 'L' terminal in a socket-outlet which is judged to be midway in the ring circuit. Call this reading *B*. This part of the test involves the use of a long test lead.

C Measure the resistance of the long test lead.

D Check that the readings $A/4 = B - C$. If both sides of the equation are virtually equal then the continuity of the phase conductor is satisfactory.

E Repeat the test for the neutral conductor and the CPC.

Example: Reading A = 0.4 ohm
Reading B = 0.2 ohm
Reading C = 0.1 ohm
Therefore $0.4/4 = 0.2 - 0.1$
$= 0.1$ ohm

Continuity of protective conductors

This test is required to verify that the conductors are both correctly connected and electrically sound. The test is carried out by using a voltage not exceeding 50 V and passing a current not exceeding 25 A into the conductor. This test simulates an earth-fault current and the level of current will show up any defects in the conductor and its terminations.

The resistance of a CPC can also be measured using a milliohmmeter and the reading obtained compared with the design resistance of the CPC, as used in the formula:

$$Z_s = Z_e + (R_1 + R_2) \text{ ohm}$$

where R_2 refers to the CPC. The value of R_2 can

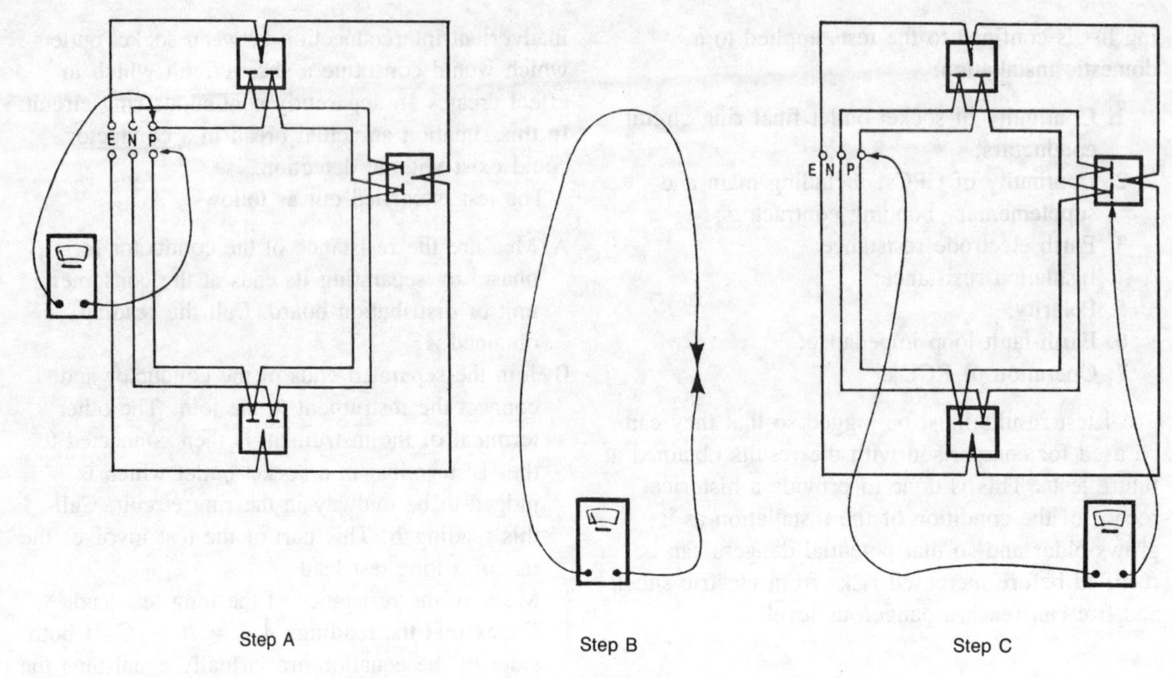

Figure 18.1. The test for continuity of ring circuit conductors.

be calculated using the figures in the relevant tables in the Wiring Regulations. These tables give the resistance of different sizes of conductors in milliohms/metre at 20 °C. If one knows the length of the CPC the total resistance can be calculated. This figure has to be used with a multiplying factor to take account of the fact that when the CPC carries a fault current heat will be generated, which will increase the CPC resistance.

Earth electrode resistance

Where the earthing arrangement is TT, with the client depending on an independent earth electrode in the ground for protection against earth-leakage currents, the testing instrument is a proprietary earth tester; or the test can be made using an extra-low voltage source.

After the earth electrode has been installed, the test ensures that the resistance of the electrode has not increased the level of earth-fault loop impedance to an unacceptable level.

Insulation resistance

There are three tests under this heading: insulation resistance (IR) between conductors; IR between conductors and earth; and the separate IR test to be carried out on disconnected equipment such as cookers.

Before these tests are carried out, it is essential to disconnect any neons and capacitors from the circuit because they will upset the readings obtained. In addition, any control devices which contain semiconductor components must also be disconnected as they can be damaged by the test voltage.

IR between conductors. For this test, all lamps have to be removed and all switches closed. The test instrument is then connected between phase and neutral (if the supply is single-phase) and the test voltage applied. The minimum acceptable value of IR is 1 megohm. In new and otherwise electrically healthy circuits, the reading will

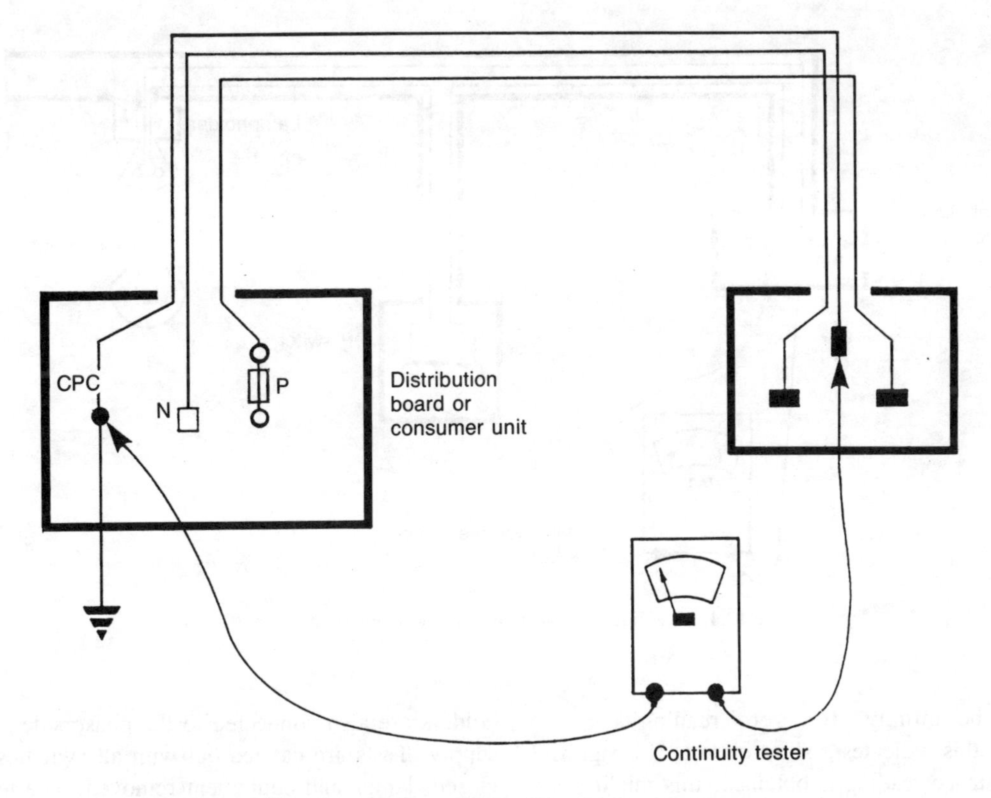

Figure 18.2. The test for continuity of the circuit protective conductor.

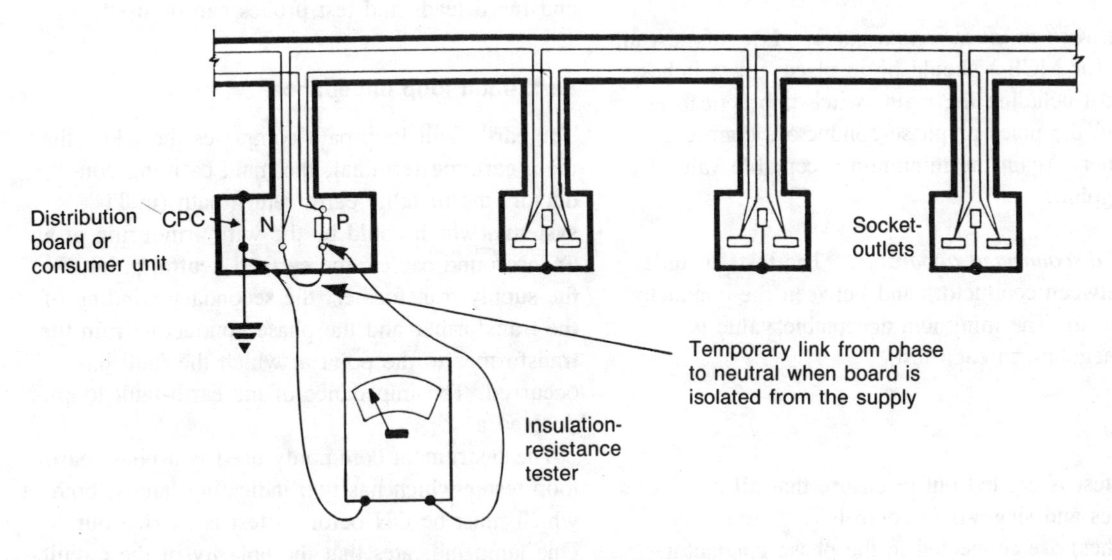

Figure 18.3. The test for insulation-resistance to earth.

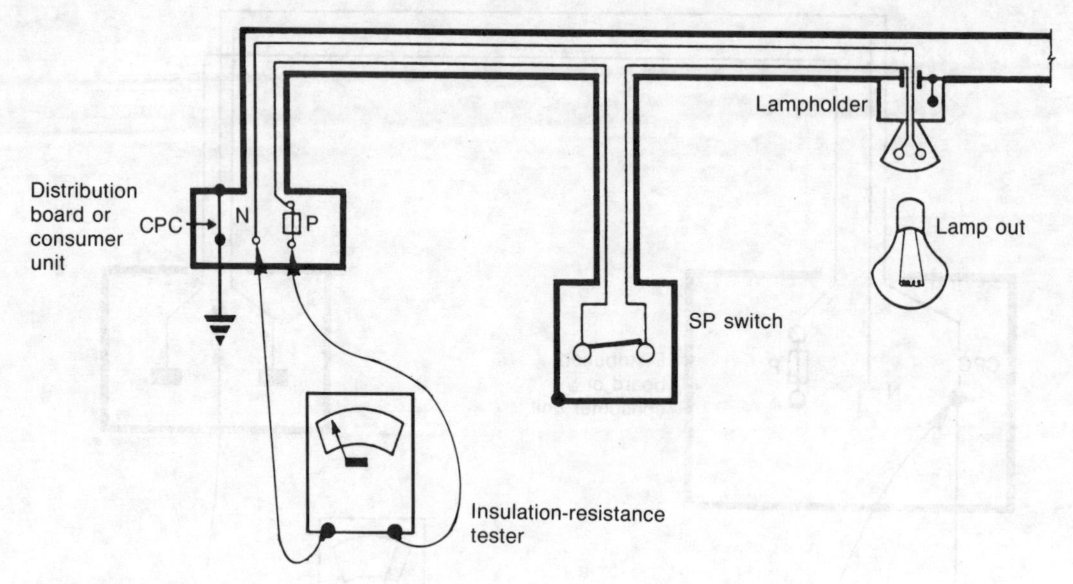

Figure 18.4. The test for insulation-resistance between conductors.

normally be 'infinity'. If a 'zero' reading is obtained, this indicates a short-circuit. If a significant resistance reading is obtained, this might indicate, say, a lamp left in the circuit (e.g. the filament resistance of a 100 W lamp would be in the region of 600 ohms).

IR between conductors and earth. For this test all fuses (or MCBs) should be in place, all switches closed (including the main switch if practicable) and all the poles or phase conductors connected together. Again the minimum acceptable value is 1 megohm.

IR of disconnected apparatus. The tests include IR between conductors and between the conductors and earth. The minimum acceptable value is 0.5 megohm, in each test.

Polarity

This test is carried out to ensure that all protective devices and single-pole controls (e.g. one-way switches) are connected in the phase conductor only. In addition the test confirms that socket-outlet 'L' terminals are connected to the phase conductor and that the centre contact of ES lampholders are also connected to the phase side of the supply. Tests are carried out with all switches closed, lamps and equipment removed. The test instrument can be a continuity tester, a low-reading ohmmeter if no mains supply is available. If the circuit is 'live' a test lamp with approved and fused leads and test probes can be used.

Earth-fault loop impedance

The earth-fault loop path comprises the CPC, the main earthing terminal, the main earthing conductor, the metallic earth return path (in TN systems, which could be the wire armouring of an underground cable), the earthed neutral point of the supply transformer, the secondary winding of the transformer and the phase conductor from the transformer to the point at which the fault has occurred. The impedance of the earth-fault loop is denoted as Z_s.

The instrument commonly used is a phase-earth loop tester which has two indicating lamps, both of which must be ON before a test is carried out. One lamp indicates that the polarity of the circuit is correct, 'L–N', and the other ensures that a proven earth connection is available, 'L–E'. There is usually a recommendation that there should be

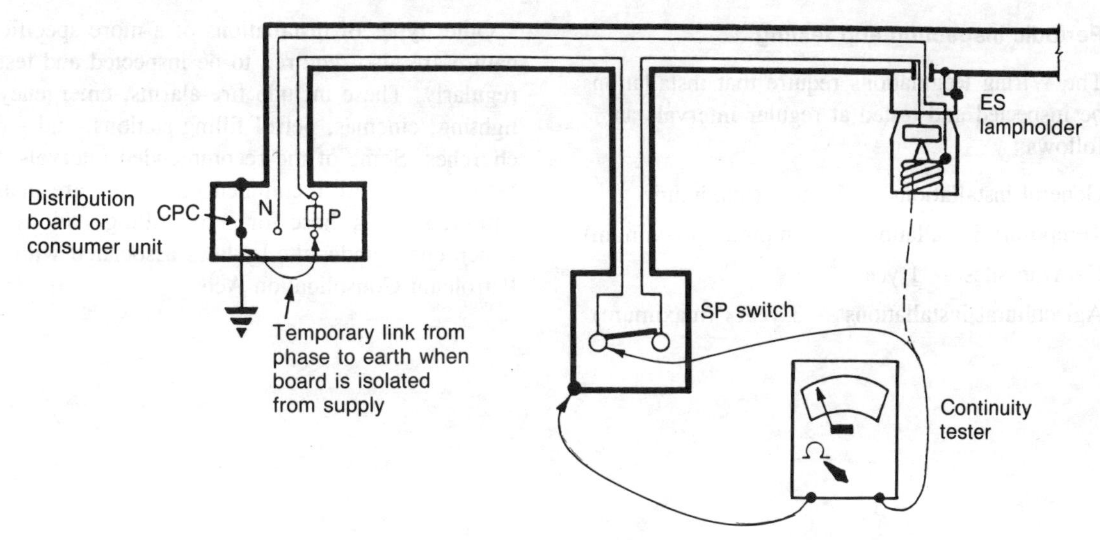

Figure 18.5. The test for correct polarity on Edison screw lampholder.

at least a 20-second interval between tests. This is to allow any heat generated in the current limiting resistor to dissipate. The instrument passes a current of around 20 A into the earth loop path.

The reading obtained must be compared with the maximum Z_s values given in Chapter 41 of the Wiring Regulations. These values vary according to (*a*) the type of overcurrent protective device and its current rating and (*b*) whether the circuit feeds socket-outlets or fixed equipment. If the reading obtained is less than the maximum value of Z_s the circuit is acceptable. If it exceeds the recommended figure, the circuit must be investigated to find the reason. The reading obtained should also be compared with the reading of the previous test (if this information is available) to see whether the Z_s value is on the increase, which might indicate a potential dangerous condition appearing in the circuit. For example, if the maximum Z_s was 2 ohms and the test reading was 1.2 ohms on a previous test but is now 1.8 ohms, this indicates that the next test would produce an unacceptable reading with the possibility that an earth fault occurring in the circuit would not produce enough fault current to operate the overcurrent protective device in either 0.4 second or 5 seconds, thus increasing the risk of electric shock to persons using the installation.

RCD tests

These devices are provided with a test button which gives the client the opportunity to check that the RCD is operational at frequent intervals. The Wiring Regulations require that an RCD be tested so that it trips within the required time. An RCD with a rated tripping current of 30 mA should trip within 0.2 second. RCD testers are available which give a choice of tripping currents, typically ranging from 5 mA to 500 mA, and which also display the actual tripping time. The higher the value of tripping current, the less becomes the time of operation. It should be noted that before the test is applied, all loads normally supplied through the RCD should be disconnected.

Certificates

The Wiring Regulations require that following the inspection and testing of an installation, completion and inspection certificates should be signed by, or on behalf of, the contractor and given to the client. These certificates then form part of the documentation relating to the installation and must be available to the person conducting subsequent tests on the installation.

Periodic inspection and testing

The Wiring Regulations require that installations be inspected and tested at regular intervals as follows:

General installations — 5 years (maximum)

Temporary installations — 3 months (maximum)

Caravan sites — 1 year

Agricultural installations — 3 years (maximum)

Other types of installations of a more specific nature are also required to be inspected and tested regularly. These include fire alarms, emergency lighting, cinemas, petrol filling stations, and churches. Some of the recommended intervals between tests are recommended; others are mandatory, such as those for petrol filling stations which come under the by-laws associated with the Petroleum Consolidation Acts.

19 Temporary installations

A temporary installation is an installation which is designed to be in service for not more than three months. If, however, the installation is required to be in service for longer than three months, it must be subjected to periodic inspection and testing at intervals of not more than three months until the installation is no longer required. Temporary installations are most often found on building and construction sites. But they could be found inside a building, for example a community hall where an exhibition is mounted and which requires electrical supplies.

So far as the Wiring Regulations are concerned, there is no distinction between a temporary and a permanent installation. In addition the requirements of the Electricity at Work Regulations apply to such installations. Further detailed information is contained in BS 7375 *Distribution of Electricity on Construction and Building Sites*.

Electricity on large construction sites is used for driving plant, welding, tools, general and security lighting, and installations in site offices, canteens and workshops. The main concern is for safety. In 1990, of the 98 employees killed on such sites, 3 died from contact with electricity or an electrical discharge. Such contact was the cause of 68 major injuries and 119 of injuries which required an absence from work for over three days.

Under the Health and Safety Act, the person on whom the responsibility falls is the 'occupier', who, on a building site, is in effect the main contractor. He, or a designated agent, is responsible for the safety of all persons on the site. This means that he has to ensure that any plant is maintained in a safe condition. Some responsibility is shifted onto the shoulders of a subcontractor, for example an electrical contractor, particularly if the latter is responsible for installing and maintaining the temperory installation for site supplies.

Usually on a large site a qualified electrician is appointed as the 'authorised' or 'competent' person to install and maintain the electrical equipment needed to distribute supplies throughout the site. Nowadays this equipment is specially designed for such use, from the supply intake unit to the extension outlet units. It is the responsibility of the electrician not to connect to the outlet units any electrical equipment which he suspects is not in a

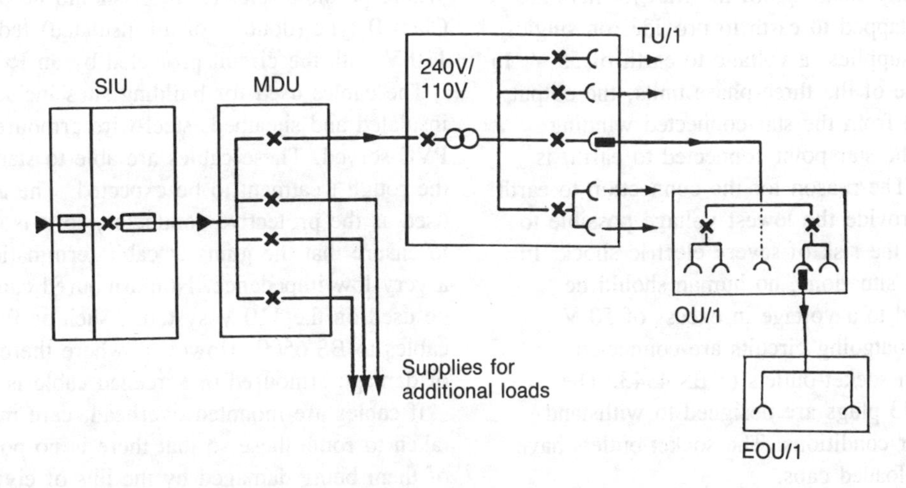

Figure 19.1. Layout for electricity distribution on a construction site.

good condition, or is clearly unsuitable for connection, for example mains-voltage non-weatherproof equipment which is intended to be used in the open in wet weather.

The main equipment recommended for installation on a building site includes:

1. *Supply intake unit.* This unit takes in a supply, usually at 415 V, three-phase. Often, however, the supply authority prefers to install the supply intake in a site hut and this is nowadays the most common provision. The main control for the whole installation is either a circuit-breaker or a switch-fuse (where the fuses are carried on the moving contact blades of the switch).
2. *Mains distribution unit.* This unit has an isolator which controls the supply to a set of busbars feeding 415/240 V supplies through moulded-case circuit-breakers (MCCBs). These offer overcurrent protection for such circuits as concrete batching machines, cranes, lifts, fixed floodlighting and supplies to site buildings, offices, canteens and the supplies to the transformer units.
3. *Transformer units.* These reduce the mains voltages down from either 415 V to 110 V, or 240 V to 110 V. Their incoming side is protected by an MCB (miniature circuit-breaker) with the outgoing circuits also protected from overcurrent by MCBs. The secondary windings of the transformers are centre-tapped to earth to provide, on single-phase supplies, a voltage to earth of 55 V. In the case of the three-phase units, the output voltage from the star-connected winding (with the star point connected to earth) is 64 V. The reason for the connection to earth is to provide the lowest voltage possible to reduce the risk of severe electric shock. In normal situations, no human should be exposed to a voltage in excess of 50 V.

   The outgoing circuits are connected through socket-outlets to BS 4343. The BS 4343 plugs are designed to withstand outdoor conditions. The socket-outlets have spring-loaded caps.
4. *Outlet units.* These have MCB protection and provide 110 V supplies to feed equipment handled by site workers.
5. *Extension outlet units.* These function like the outlet units but have no MCB protection incorporated. They are mainly intended for connection to a single- or three-phase source of supply.

Colour identification is used to denote the voltages available at socket-outlets: yellow = 110 V; blue = 240 V. Other voltages which may be found on a building site include: 25 V (violet); 50 V (white); 415 V (red); over 415 V (black).

The plugs and socket-outlets are so designed that they cannot be interchanged, so preventing, say, a 110 V plug being inserted into a 240 V socket-outlet.

Site lighting is often a requirement when the hours of darkness curtail the working time on site. The lighting is often provided by means of festoon cables fitted with BC or ES lampholders. The lighting cables are available in 50 and 100 m lengths and are provided with festoon wire guards. All handlamps are fitted with wire guards.

While much of the electrical equipment on a site is rated for 110 V, lower voltages are used for situations where water is present. Handlamps, for instance, should be supplied from a 50 V transformer with its secondary winding centre-tapped to earth thus reducing the shock voltage to 25 V. Where possible, electric tools should be of the Class II type (double- or all-insulated) fed from 110 V with the circuit protected by an RCD.

The cables used for building sites include PVC-insulated and sheathed, steel-wire armoured and PVC-served. These cables are able to stand up to the rough treatment to be expected. The armour is used as the protective conductor, so it is essential to ensure that the joints at cable terminations have a very low impedance. Non-armoured cables can be used on the 110 V systems, such as flexible cables to BS 6500. However, where there is a risk of damage armoured or screened cable is better.

If cables are mounted overhead, care must be taken to route these so that there is no possibility of them being damaged by the jibs of civil engineering plant. Barriers are required to keep mobile plant at a safe distance.

No temporary installation will remain safe for long if it is not subjected to regular maintenance. The maximum interval between inspections and tests is three months. But it is up to the 'competent' person to assess whether reduced periods are required, for instance weekly. All hand-held equipment must be in a safe electrical condition before it is used. In short, a complete record of all equipment must be maintained and defective units taken out of service until repaired. The 'competent' person should not be one of the electricians engaged in work on the site. He or she should be given the responsibility and that duty should be his or her only concern.

In addition to BS 7375, further guidance for 'competent' persons is given in two Guidance Notes issued by the Health and Safety Executive: GS 24 *Electricity on Construction sites* and PM 32 *The Safe Use of Electrical Apparatus*.

While this chapter is concerned with 'temporary' installations on building sites, 'temporary' can also be applied to other situations. For example, in a factory temporary connections have to be made to ensure continuity of supply while, perhaps, a machine is taken out of service. It cannot be stressed too much that certain precautions must be taken.

Before making temporary connections, the supply must be switched off and locked off. When the connection has been made, it is important to ensure that the earthing system is restored correctly in both the existing installation and the temporary circuit. Again, when taping or temporarily sealing off cable ends, extra care must be taken that the cable cores and the protective conductors are separated and separately sheathed.

Temporary supplies to machines must be fully protected and runs secured properly. These supply circuits must also be included in the existing arrangements in the workshop provided for isolation and emergency stopping.

20 Circuit-control devices

All electrical circuits are required to have some means whereby they can be energised and disconnected from their supply source. This is done by switches, of which there is a very wide variety of types available. A 'switch' is defined as a mechanical device capable of making, carrying and breaking current under normal circuit conditions, which may include specified overload conditions. Switches in domestic installations are familiar devices used to control the supply to lighting, cooker and water-heating circuits. Socket-outlets may have switches incorporated. In a consumer unit, the main switch isolates the whole installation from the supply.

Certain types of circuit controls do not qualify as switches. These include thermostats for water-heaters and heating equipment, and touch switches, or electronic switches. Some switches are used as isolators which are designed to disconnect a circuit usually when the circuit has no current flowing in it.

Some switches are operated by an electromagnet; these include contactors used for switching heating loads, large lighting loads and are also incorporated in motor starters. A more specialised type of electromagnet-operated device is the relay.

Although circuit-breakers tend to be regarded as devices used for protection of circuits against overcurrent (overload and short-circuit), they also perform a duty as switches.

Circuit conditions

Every electrical circuit has its own characteristics, which means that it will show some peculiar electrical property depending on the type of load connected to it. For instance, a circuit which has a purely resistive load (a resistor used as a lamp filament, or heater element) will show a current which rises when the circuit is first switched on and then falls as the element reaches its normal operating condition. This means that the switch or other circuit-control device must at least be able to break the full-load current taken by the resistor. This applies particularly if the circuit has a dc supply. If, however, the supply is ac, when the switch contacts separate there may be a small arc drawn out between the contacts. This characteristic is even more noticeable when the resistor is in the form of a coil (e.g. in a firebar element). This effect is caused by the electrical property which a coil has in an ac circuit. It is called the 'inductive effect' and is explained more fully in Chapter 24.

If, instead of a resistive conductor wound in the form of a coil, a low-resistance conductor is wound round a soft-iron core, the item is then known as a 'choke' or inductor, and the circuit is said to have 'inductive characteristics', which lead to switching problems. A fluorescent circuit is an inductive circuit, as is a motor circuit.

If the circuit has a capacitor included in it, it will also show certain characteristics which may be shown as arcing between switch contacts as they separate. The most pronounced effects of the inclusion of an inductor or a capacitor in a circuit is seen when an ac supply is used. However, small capacitors are often used connected across switch contacts to absorb the sparking caused by contact separation. Used in this way they are sometimes called 'radio-interference suppressors' (e.g. in fluorescent lamp switch starters).

Thus, before a circuit-control device is chosen the circuit to be controlled must be studied so that the device can handle, without damage to itself or the associated circuit wiring, the conditions in the circuit when it is connected or disconnected from its supply. The sections in this chapter which follow indicate the type of control for a circuit which various devices offer.

Contacts

There is in existence an extremely wide range of electrical-contact types used to control the flow of an electric current in a circuit. The action of any pair or pairs of contacts is (*a*) to 'make', to allow the current to flow, and (*b*) to 'break', to prevent the current flow. When this action is contained in a specially-designed wiring accessory or apparatus it becomes one of the many forms of devices used to control circuits: switches, contactors, circuit-breakers and the like.

The basic requirements of any pair of contacts are (*a*) low resistance of the contact material and (*b*) low resistance between the two contact surfaces when they meet to make the circuit.

When these requirements are satisfied, the two main factors which lead to switch troubles are very much reduced. Though one can choose a low-resistance contact material (e.g. copper), one cannot always control the amount of pressure required to keep the two contact surfaces closed sufficiently to reduce what is called 'contact resistance'. A switch, for instance, which is operated many times will eventually reach a state when its springs become weakened, with the result that pressure of the contacts is lost to such an extent that heat is generated and a breakdown of the switch follows.

The higher the resistance of contact material the more heat (I^2R watts) there will be when a current passes along it. The second factor involved in the design of switch contacts is the amount of pressure needed to keep the two contact surfaces together. All circuit-control devices which meet the relevant specifications of the BSI are tested very rigorously to ensure that they stand up to more wear and tear than they would meet with in normal use. Even so, most contact troubles met with in practice involving the use of circuit-control devices can be traced to insufficient contact pressures.

The material most often used for contacts is copper; this is because it is available in commercial quantities and it has a very low resistance. The terminals associated with the contacts, to which cables and wires are attached, are most often made from brass or phosphor-bronze. These two metals are much harder than copper and so can withstand a certain amount of rough handling with screwdrivers when wiring is being carried out.

The insulating materials used in circuit-control devices include vitrified ceramic (for the bases of switches), bakelite (for switch covers and cases), nylon and mica (for carrying the moving contacts of switches), and insulating oil (used in oil-break circuit-breakers).

In many circuit-control devices silver is used, either as a contact facing, or as the contact itself. The material has a resistance lower than that of copper; it also has high heat-dissipation characteristics and is, for this application, economical to use. Motor-control switches sometimes have contacts of silver-cadmium oxide to reduce the tendency to weld together with heat.

Liquid mercury is also used in special switches called mercury switches. This material has a low contact resistance and a high load-carrying capacity, and can be used in situations with ambient temperatures from about −17 to 204 °C.

Because the contacts are the heart of the circuit-control device, it follows that their surfaces must be kept clean at all times. Cleaning fluids are available for this purpose. Other maintenance points are the periodic tightening up of conductor terminals and connections, and ensuring that springs have not weakened through use, or that cam surfaces have not become worn.

There are two classes of duty for circuit-control devices: (*a*) light current and (*b*) heavy current. Into the first class fall generally lighting switches, relays and bell-pushes; the second class includes contactors and circuit-breakers.

Switches and switchfuses

N.B. Circuits mentioned in this section are detailed in Chapter 27.

A switch is a device for controlling a circuit or part of a circuit. The control function consists of energising an electrical circuit, or in isolating it from the supply. The type of switch generally indicates the form which this control takes. For instance, a single-pole switch (usually called 'one-way') controls the live pole of a supply. A double-pole switch controls two poles.

A common type of switch in use today is the micro-gap with a rating of 5 A, to control lighting circuits. Switches with a 15 A rating are also used to control circuits which carry heavier currents on both power (socket-outlet) and lighting arrangements.

Switches are designed for use on dc and/or ac. In a dc circuit, when the switch contacts separate, an arc tends to be drawn out between the separating surfaces. This arc is extinguished only when the contacts are far enough apart and when the breaking movement is quick.

Investigation of a dc switch will indicate the length of the gap required when the switch is open. Compare this gap with the gap length on an ac-only switch; it will be found that the latter is very much smaller. The reason for this is that ac tends to be what is called 'self-extinguishing'. In an ac circuit, during the time taken for the contacts to open, the voltage, which is alternating, varies between zero and a maximum. It is at the zero position of the alternating voltage that the arc drawn between the parting contacts of an ac-only switch is extinguished — and it does not establish itself again in normal circuit conditions. Thus, a switch designed for use only on an ac system need have only a small gap and, furthermore, the contact movement does not require to be operated so rapidly as is the case with dc switches.

Quick-make-and-break switches are used for dc circuits. Quick-make, slow-break switches are recommended for ac circuits, particularly where the load is an inductive one, for instance where fluorescent lamps are being used.

The most common lighting circuits are controlled by using one-way and two-way switches, double-pole switches and intermediate switches. Other types of circuit-control devices and switches are dealt with later in this chapter.

The single-pole, one-way switch provides the ON and OFF control of a circuit — from one position only. When the switch is closed, the lamp is on; when the switch is open, the lamp is off. One-way switches are mounted with the word 'TOP', which appears on the back of the switch plate, at the top. This is to ensure that when the switch rocker is in the up position, the circuit is disconnected from the supply. The switch is, of course, connected in the phase conductor only.

The double-pole switch is used in any situation where the voltage of the neutral conductor of a supply system is likely to rise an appreciable amount above earth potential: use of the double-pole switch means that a two-wire circuit can be completely isolated from the supply. The usual application is for the main control of sub-circuits and for the local control of cookers, water-heaters, wall-mounted radiators, and other fixed current-using apparatus. The double-pole switch is often used for the 'master' control of circuits, the switch being operated by a 'secret key' attachment, and in consumer units for the complete isolation of an electrical installation from the supply.

The two-way switch is basically a single-pole changeover switch offering two alternative routes for the passage of the circuit current. These switches are sometimes known as 'landing' switches from the days when their application in the electrical installation was virtually limited to 'one in the hall, and one on the landing upstairs'.

Though the two-way switch is still used extensively for stair lighting, it is also to be found wherever it is necessary to have one or more lights controlled from any one of two positions. They are nowadays to be found in bedrooms (door and bedside), long halls (at each end) and particularly in any room with two entry doors (one at each door).

In design, the switch has four terminals, two of which are permanently connected together inside the switch by a small copper bar on what is called the 'bar' side. One of the bar terminals is blanked off to form a non-separable contact. The switch feed is taken to the other open terminal on the bar side. The two other terminals are connected to the 'strapping wires'. Two-way switches are used in pairs, interconnected so that the switchwire of the light circuit is taken from the open terminal on the bar side of the second switch.

The intermediate switch offers control of a circuit from any one of three positions, the other two positions being at the two two-way switches with which the intermediate switch is most often used. The intermediate wiring circuit is basically a two-way circuit in which the strapping wires are cross-connected by the two ON positions of the intermediate switch. There are two different kinds of intermediate switch, one of which is in common use. It is thus advisable to check the type with an ohmmeter, or bell-and-battery set, because the

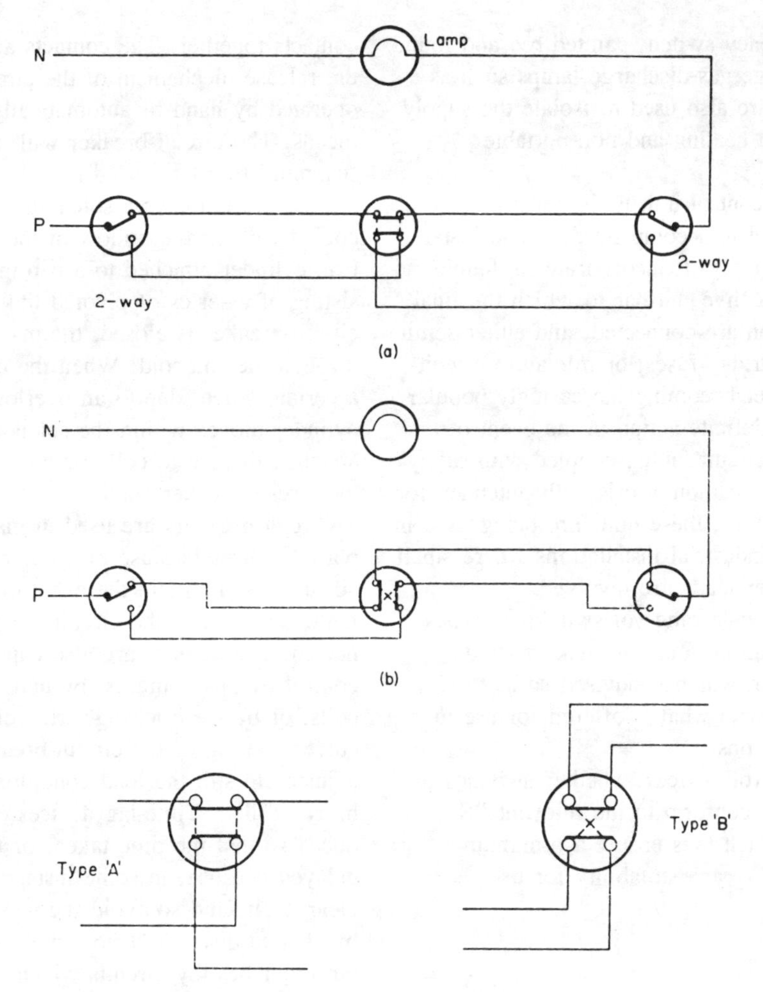

Figure 20.1. (a) and (b) Wiring diagrams for intermediate switching circuits.

method of connecting up differs. Figure 20.1 shows the two common forms of connection made within each type of switch.

The application of the intermediate switch in electrical installations has so far been very limited. But there is no reason why it should not be used more extensively. Long halls, corridors and passage-ways with many doors are still wired up for two-way control. For reasonable convenience the light or lights should be controlled from every door and entrance. Thus, the user of this type of circuit can make his way through a house, switching on lights before him, and switching off behind him without have to grope about in the dark.

Two or more intermediate switches can be interconnected into the basic two-way circuit to offer control from an almost unlimited number of positions.

The switchfuse is often found as the 'main switch', near the supply-intake position. It is a unit in which the main switch (for installation control) and the main fuses (for the protection of the installation) are combined. In all instances, the switch of the switchfuse cannot be operated when the cover is open, nor can the cover be removed or opened while the switchfuse is closed. The switchfuse, which usually controls a separate distribution board, is of the double- or triple-pole type, depending on the supply system.

Double- and triple-pole switches are found in metal-clad units called isolators. An example is the fireman's emergency switch, painted red and found beside high-voltage gas-discharge lamps such as neons. Isolators are also used to isolate the supply from motors, and heating and non-portable appliances.

The consumer control unit is the most common means used to isolate a complete domestic installation from the supply. It incorporates a double-pole switch and a 'live' busbar to which the final circuits' protection are connected, and either semi-closed fuses, cartridge fuses, or miniature circuit-breakers, the latter becoming increasingly popular because of their definite action in the event of overloading and circuit faults, coupled with safety in their operation. Although originally intended for domestic installations, these units are being used in commercial and industrial installations where small lighting and power loads are involved.

The extremely wide range of switchgear types available today can be found in makers' trade literature, study of which is advised so as to become familiar with what is offered for use in electrical installations.

All circuit-control devices, whether switches or other types, must conform to the relevant BS specifications, which thus ensure a minimum guarantee of quality and suitability for use.

Circuit-breakers

The circuit-breakers can be regarded as a switch which can be opened automatically by means of a 'tripping' device. It is, however, more than this.

Whereas a switch is capable of making and breaking a current not greatly in excess of its rated normal current, the circuit-breaker can make and break a circuit, particularly in abnormal conditions such as the occasion of a short-circuit in an installation. It thus disconnects automatically a faulty circuit.

A circuit-breaker is selected for a particular duty, taking into consideration the following: (*a*) the normal current it will have to carry and (*b*) the amount of current which the supply will feed into the circuit fault, which current the circuit-breaker will have to interrupt without damage to itself.

The circuit-breaker generally has a mechanism which, when in the closed position, holds the contacts together. The contacts are separated when the release mechanism of the circuit-breaker is operated by hand or automatically by magnetic means. The circuit-breaker with magnetic 'tripping' (the term used to indicate the opening of the device) employs a solenoid which is an air-cooled coil. In the hollow of the coil is located an iron cylinder attached to a trip mechanism consisting of a series of pivoted links. When the circuit-breaker is closed, the main current passes through the solenoid. When the circuit rises above a certain value (due to an overload or a fault), the cylinder moves within the solenoid to cause the attached linkage to collapse and, in turn, separate the circuit-breaker contacts.

Circuit-breakers are used in many installations in place of fuses because of a number of definite advantages. First, in the event of an overload or fault, all poles of the circuit are positively disconnected. The devices are also capable of remote control by push-buttons, by under-voltage release coils, or by earth-leakage trip coils. The overcurrent setting of the circuit-breakers can be adjusted to suit the load conditions of the circuit to be controlled. Time-lag devices can also be introduced so that the time taken for tripping can be delayed because, in some instances, a fault can clear itself, and so avoid the need for a circuit-breaker to disconnect not only the faulty circuit, but other healthy circuits which may be associated with it. The time-lag facility is also useful in motor circuits, to allow the circuit-breaker to stay closed while the motor takes the high initial starting current during the run-up to attain its normal speed. After they have tripped, circuit-breakers can be closed immediately without loss of time. Circuit-breaker contacts separate either in air or in insulating oil.

In certain circumstances, circuit-breakers must be used with 'back-up' protection, which involves the provision of HBC (high breaking capacity) fuses in the main circuit-breaker circuit. In this instance, an extremely heavy overcurrent, such as is caused by a short circuit, is handled by the fuses, to leave the circuit-breaker to deal with the overcurrents caused by overloads.

In increasing use for modern electrical installations is the miniature circuit-breaker (MCB). It is used as an alternative to the fuse, and has certain advantages: it can be reset or reclosed easily; it gives a close degree of small overcurrent protection (the tripping factor is 1.1); it will trip on a small sustained overcurrent, but not on a harmless transient overcurrent such as a switching surge. For all applications the MCB tends to give much better overall protection against both fire and shock risks than can be obtained with the use of normal HBC or rewirable fuses. Miniature circuit-breakers are available in distribution-board units for final circuit protection.

One main disadvantage of the MCB is the initial cost, although it has the long-term advantage. There is also a tendency for the tripping mechanism to stick or become sluggish in operation after long periods of inaction. It is recommended that the MCB be tripped at frequent intervals to 'ease the springs' and so ensure that it performs its prescribed duty with no damage either to itself or to the circuit it protects.

The circuit-breaker principle is used for protection against earth leakage and is discussed more fully in Chapter 16.

Contactor

When a switching device has one or more switches in the form of pivoted contact arms which are actuated automatically by an electromagnet, the device is known as a contactor. The coil of the electromagnet is energised by a small current which is just sufficient to hold the pivoted contact arm against the magnet core, and in turn so hold the contacts (fixed and moving) together. Contactors are used in an extremely wide range of applications.

They fall into two general types: (*a*) 'maintained' and (*b*) 'latched-in'. In the first type, the contact arm is maintained in position by the electromagnet. In the latched-in type, the contact arm is retained in the closed position by mechanical means.

Contact design and material depend on the size, rating and application of the contactor. Contactors with double-break contacts usually have silver cadmium-oxide contacts to provide low contact-resistance, improve arc interruption and anti-

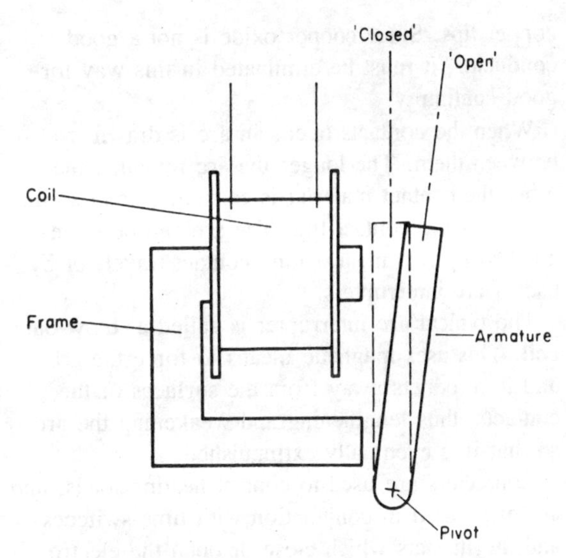

Figure 20.2. Typical contactor arrangement.

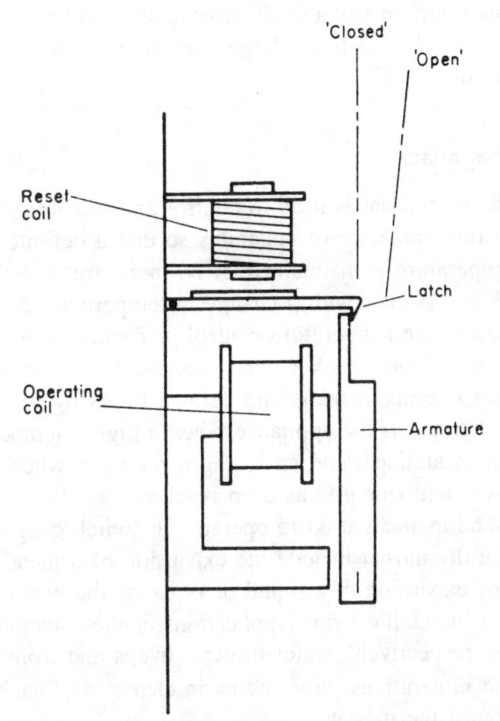

Figure 20.3. Typical latched-in contactor arrangement.

welding characteristics. Large contactors with single-break contacts use copper contacts for economy. Usually single-break contacts are designed with a wiping action to remove the copper-oxide film which readily forms on the copper tips. Since copper oxide is not a good conductor, it must be eliminated in this way for good continuity.

When the contacts open, an arc is drawn between them. The longer the arc remains, the more the contact material is consumed, and so the shorter is the contact life. The arc can be extinguished by two means: long contact travel, or by use of arc interrupters.

The typical arc interrupter is called a 'blow-out' coil. This uses magnetic means to force the arc and its products away from the surfaces of the contacts, thus lengthening and weakening the arc so that it is eventually extinguished.

Contactors are used to control heating loads, and are often used in conjunction with time-switches and thermostats which close or open the electromagnet current as required. With the contactor, a small current (for the electromagnet) can be used to control a relatively large current in another circuit.

Thermostat

The thermostat is used to control an electric heating appliance or apparatus so that a definite temperature is maintained. It is, therefore, a switch which operates with a change in temperature and is used in the temperature control of rooms, water-heaters, irons, cooker ovens and toasters. It maintains a temperature within defined limits by switching off the appliance when a higher temperature is attained, and switching it on again when a lower temperature has been reached.

The methods used to operate the switch contacts of a thermostat include the expansion of a metal rod, expansion of a liquid or a gas or the bending of a bimetallic strip. Applications of these methods are, respectively, water-heaters, ovens and irons. The illustrations show the basic elements of each type of thermostat.

The speed of response of a thermostat to a change in temperature depends to a large extent on the material used to convey the heat, called the controller. A thermostat whose thermally-sensitive elements are directly opposed to the heat transfer medium will respond faster than one whose elements are shielded by a housing. Liquid-filled systems respond more quickly than gas-filled systems.

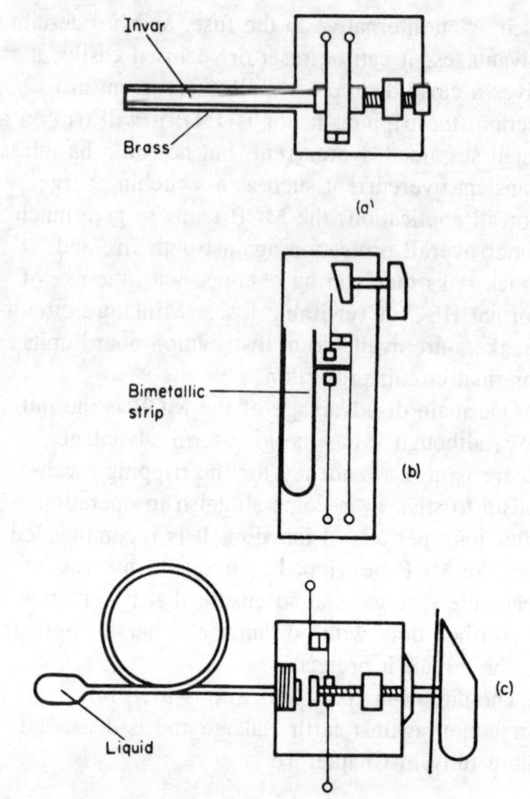

Figure 20.4. (a) Thermostat with metal rod; (b) Thermostat with bimetallic strip; (c) Thermostat with liquid bulb.

Simmerstat (energy controller)

The simmerstat operates by opening and closing a switch at definite time intervals. The ratio of the time the controlled circuit is 'on' to the time it is 'off' is determined by placing a graduated knob at a particular setting. The simmerstat provides a gradual and infinitely variable means of control of, for instance, a cooker hotplate.

The simmerstat consists of a bimetallic strip surrounded by a heating coil. There are two contacts.

Refer now to Figure 20.5. A control knob operates a cam which varies the setting of the device. When the control knob is in the OFF position, the cam depresses the free end of the bimetallic strip which is in the form of a U,

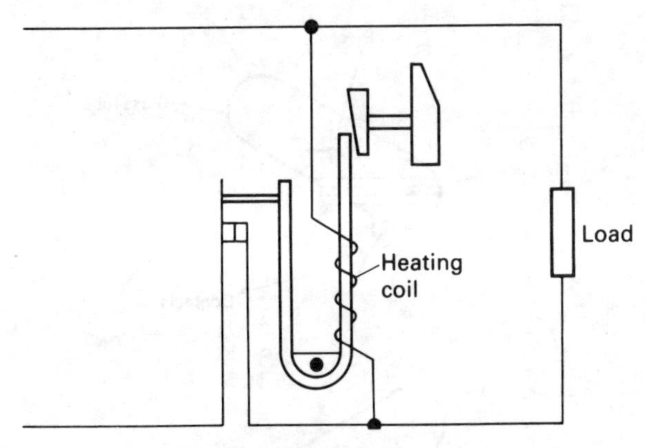

Figure 20.5. Simmerstat.

pivoted at the bend. The depression causes the two contacts to open. With the control knob in an intermediate position, the cam allows the free end of the strip to rise and the contacts to close. This completes the circuit of the hotplate and also the heat coil of the simmerstat. After a time, the heat produced by the heat coil causes the bimetallic strip to bend in such a way that the ends of the U move apart. The upper end meets the cam. As it cannot go any further, the lower end continues to take up the movement to travel downwards and so open the contacts. The supply to the hotplate, and to the heat coil, is thus interrupted and will remain so until the strip cools sufficiently to allow the contacts to close again to repeat this cycle.

When the control knob is placed in the ON position, the cam allows the strip to rise freely to the extent that the contacts are never opened. The hotplate now uses its full rated power.

Special switches

With the extensive use of electricity today, it is not surprising to find that there is a great variety of switches and other circuit-control devices with special applications. It is possible to indicate here only some of the most common types.

Three-heat switch

This type of switch is most often associated with the grill-plate of an electric cooker, though it is also used for the heat control of boiling plates. The circuit controlled by the switch consists of two elements of equal resistance. The three-heat switch then offers low, medium and high heat values by its three positions.

Figure 20.6 shows the connections. For low heat, both elements are connected in series to give

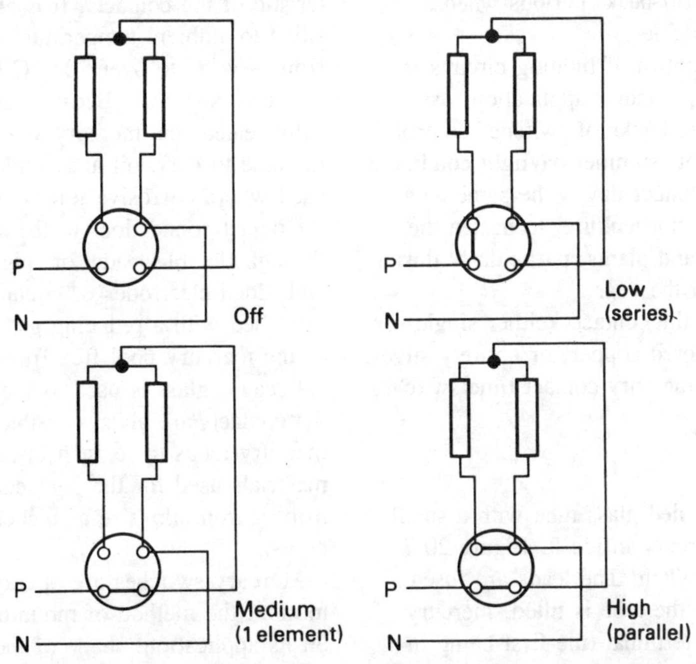

Figure 20.6. Circuit connections for a three-heat, series-parallel switch.

25 per cent available power. For medium heat, only one element is connected to give 50 per cent power. For the high-heat condition (full power) the elements are connected in parallel.

The three-heat switch is essentially a rotary or turn switch. The positions are OFF, LOW, MEDIUM, HIGH. The switches are available as a single-pole type (four terminals) or a double-pole type (five terminals).

Time switch

As indicated by its definition, the time switch introduces a 'time element' into an electrical circuit, so that automatic control of the circuit is available at predetermined times. Time switches fall into two general groups: spring-driven and motor-driven. The former uses a mechanism similar to that found in clocks. The latter group uses as the driving unit a small electric (synchronous) motor whose speed is constant and varies only with the 50 Hz frequency of the mains supply. Similar motors are used in electric clocks.

There are many applications for time switches: shop-window lighting, driveway lighting, street lighting, staircase lighting in multi-tenanted buildings and heating loads, the latter being switched on during 'off-peak' periods when a cheaper tariff is available.

The time-switch control of lighting circuits is often found in such particular applications as poultry houses, where banks of switches control the lighting to simulate summer-daylight conditions and so introduce a 'longer-day'. The same technique is also used in horticulture, to hasten the growth of seedlings and plants, particularly during off-season periods of the year.

For normal work, the contacts (either single- or double-pole) are silvered copper, or entirely silver. For heavy currents, mercury-contact time switches are used.

Mercury switch

This is basically a sealed glass tube with a small amount of liquid mercury inside it. Figure 20.7 illustrates a typical switch. The leads are fused into the glass. When the tube is tilted, mercury flows over a second terminal (the first being in permanent contact with the mercury). Thus,

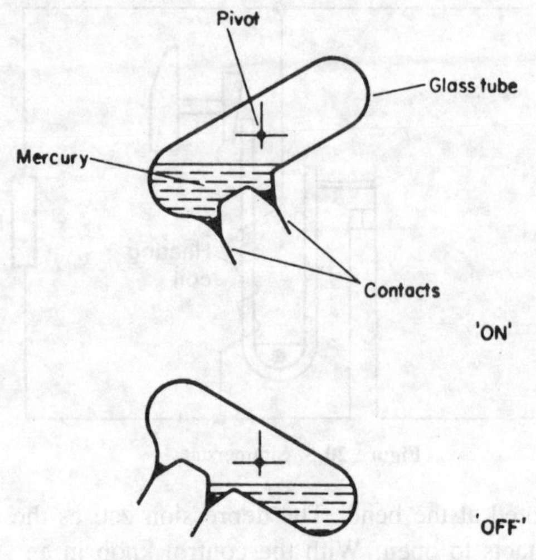

Figure 20.7. The mercury switch.

contact is made to make the circuit. Mercury switches are made in a very wide variety of types, each type being designed with a particular duty and application in mind.

Switches of this type have many advantages: low force required to operate them, low contact-resistance, high load-carrying capacity, low cost, and a long life because of the 'no wear' characteristic of the contacts. It is also relatively insensitive to ambient temperature conditions; a range from −4 °C to over 204 °C has been specified for some switches. Because the glass is hermetically sealed, the mercury switch is effectively immune to dust, oil and condensation, and can be used where corrosive fumes are present.

Contact connections to the switch are made through flexible leads, or 'pigtails', attached to the embedded electrodes or contacts. Some switches are filled with a reducing gas to keep the surface of the mercury pool free from tarnish.

Because glass is used as the switch container, the contacts are always visible for inspection; and mercury tends to resist heat and arc effects. The materials used for the contacts include tungsten, iron or iron alloys (e.g. nickel-iron) and mercury pools.

Mercury switches are operated by a tilting motion; the method of mounting a switch depends on its application, shape of the actuating member, and the motion produced by it. In the case of a single-throw switch, the glass tube is tilted from the horizontal. Mountings include bimetallic strips, cams and rotating levers. A time-lag element can be introduced by restricting the flow of mercury from one position to another; this is done by a wall placed inside the tube. The wall contains a hole, the diameter of which determines the amount of time-delay.

Rotary switch

The rotary or turn switch offers the facility of controlling a large number of circuits from a local position by using one switch. The three-heat switch is one of the most common examples of the rotary switch. Others include the switches used on switchboards in conjunction with ammeters and voltmeters on three-phase systems to indicate phase-to-phase currents and voltages.

Many banks of contacts can be fitted to a rotary switch so that complete control of circuits is available. Generally the currents are not large: 15 A is the usual limit.

Micro-gap switch

This switch derives its type name from the fact that when its contacts (usually silver) are open they are separated by an extremely small gap: anything up to 3 mm. As indicated earlier in the section on contacts, such switches can be used only on ac circuits. They have many applications apart from 'ac only' lighting circuits.

Thermostats using a 'snap-acting' bimetallic element are in effect micro-gap switches and are to be found in the temperature control of irons, toasters, and cooker heating elements. One industrial application is where a motor overheats and a bimetallic, snap-acting device will switch off the energising current to stop the motor and so protect its winding.

The snap action is always positive in these switches, no matter how rapidly or how gently the force is applied to the operating button. The button can be moved by a plunger, a leaf spring, or a roller and a lever.

Starter switch

Starter switches are used for starting fluorescent lamps. The glow-type starter switch consists of two separated bimetallic contact strips contained in a glass bulb filled with helium gas. The contacts are connected to the fluorescent lamp filaments. When the circuit-control switch is closed, the mains voltage appears across the two contact strips. This voltage is sufficient to cause a small gas discharge. The heat generated by the discharge affects the bimetallic contact strips which bend forward to meet each other. When they make contact, the current flows through the fluorescent lamp filaments to heat them. The gas-discharge glow in the starter switch now disappears. After a few seconds the bimetallic contact strips cool down and separate. This sudden interruption of the circuit causes a high-voltage surge to appear across the ends of the main lamp electrodes to start the gas discharge.

The voltage which now appears across the contact strips in the starter switch is, during running conditions, insufficient to cause further discharge in the helium gas, and so the contacts remain open while the main lamp is burning.

Two-way-and-off switch

This is a single-pole changeover switch with an OFF position. It is to be found in hotels, ships and hospitals where it is required to have two lamps in circuit while so arranging their control that both cannot be used at the same time.

The two-way-and-off switch can be used as a dimmer control, when in one ON position of the switch only one lamp is lit; in the other ON position, two lamps are connected in series to give a 'dim' light. Other lamp-control arrangements are available when this type of switch is used with other types such as the two-way.

Series-parallel switch

This is a three-position switch with an OFF position when the switch knob or dolly is central. The switch is used to control two points, or two groups of points. In one ON position, the lamp or lamps are connected in series (dim). In the other ON position, the lamp or lamps are connected in parallel (bright). These switches are to be found in hotel corridors, hospital wards and in railway carriages.

Low-voltage contacts

The most common type of low-voltage contact is the bell-push, which is operated by the direct pressure of a finger on a push-button: the contacts are copper or brass. One is fixed to the base of the bell-push, the other is fixed at one terminal end, its other free end being raised. Pressure on the push-button depresses the contact's free end to complete the circuit. The contacts are usually natural copper, though they are sometimes given a coating of non-oxidisable metal. Other low-voltage contacts use steel springs and phosphor-bronze springs, and are associated with various alarm circuits: burglar, fire, frost, water-level and smoke-density.

Relay

The most common relay is a switch operated by an electromagnet. It consists of an iron-cored coil and a pivoted armature. When the coil is energised, one end of the armature is attracted to the electromagnet and the other end presses two or more contacts together; contacts may also be opened by this movement of the armature.

Relays are either normally-closed (NC) or normally-open (NO). In the first type, when the coil is energised the contacts are open; the contacts close when the coil is de-energised. In the NO relay, the contacts are closed when the coil is energised, and open when it is de-energised. In effect, the relay is an automatic switch.

Relays are normally designed to operate when a very small current flows in the coil. Thus, a small current can be made to switch a larger current on or off, just as a contactor functions from a distant point (remote control). They are also used in bell and telephone systems, and have a wide application in industry.

Other types of relays use a solenoid for their operation. In this instance a plunger is attracted when a predetermined value of current flows in the coil. A time-lag element can be introduced by the addition of an oil- or air-dashpot to delay the movement of the plunger.

Induction and impedance relays operate by the movement of a pivoted disc in the field of an electromagnet; the protective device (usually a circuit-breaker) with which these types are associated is operated by small contacts on the moving disc which, when they close, trip the circuit-breaker. They are used in the protective systems for supply systems, motors, generators and transformers.

The thermal relay consists of a bimetallic strip which heats up when the operating or circuit current flows through it or through an adjacent heating coil. The bending of the strip causes the contacts to either make or break.

Fireman's switch

This switch is used to isolate high-voltage lighting circuits usually found on the exterior walls of buildings, such as neon signs. The switch, which is painted red, is mounted on the outside of the building adjacent to the sign lamps. A label 'Fireman's switch' is required to be mounted close to the switch. The OFF position of the switch is at the top and there must be a catch (spring-loaded) to prevent its inadvertent return to the ON position. The mounting height should be not more than 2.75 m from ground level.

Emergency switching

This is a requirement of the Wiring Regulations. The switches take the form of large mushroom-head buttons which can be knocked in the event of an emergency, say, in a workshop. The switch then disconnects the circuit or machine.

General requirements

Directly operated switches are not allowed in bathrooms or shower rooms where switches are within reach of a person in contact with the bath or shower. Pull-cord switches are recommended in these situations.

When time switches are being connected up, it is essential to ensure that a CPC is also connected to the earth terminal provided. From time to time the consumer may need to make adjustments to the switch settings, thus coming into contact with metal parts such as the switch-operating levers. Correct use of the earthing terminal will prevent shock risks.

All lighting switches must be connected in the phase conductor only and the correct colour coding of the connecting wires is required by the Wiring Regulations. Any exposed metalwork (such as a metal switch plate) must be earthed. The switch must be of an adequate current rating. If they are used for inductive loads such as fluorescent circuits, they must be fully rated for the value of inductive current taken. If they are not, then they must not carry any more than half their rating, e.g. 2.5 A in the case of a 5 A-rated switch.

Where switches are used as isolators for motor circuits, they should be located close to the motor position. If this is not possible, the switch handle should be able to be padlocked in the OFF position so that work can be carried out without fear of the circuit becoming live.

Protection (physical)

The physical protection of circuit-control devices involves the provision of some method of enclosing the contacts and the switch-movement mechanism. In most instances this is merely providing an insulating enclosure so that the device can be operated with safety from electric shock, hence 'shockproof'. Other enclosures include 'weatherproof' and 'watertight'.

For industrial installations, full protection is often given to the operating dollies of switches by a corner-fixed cover-plate which is both dished and lipped, the dolly being located within the dished part of the cover.

Dustproofing is necessary where there is the possibility of fine metal cuttings and dusts, fluff or oil-mists penetrating to the contacts.

21 Cells

An electric cell is basically two different metallic plates in a conducting liquid (electrolyte). Figure 21.1 shows a simple cell using copper and zinc plates and the direction of current flow in an electrical circuit containing the cell. It should be noted that carbon, though not a metal, is used in Leclanché cells.

The current flow is a result of the chemical energy stored within the cell being converted into electrical energy in the external circuit. All cells operate on the same basic principle and the plates may be carbon and zinc, as in the Leclanché cells, or two lead plates each having a different active material on their surface as in an accumulator.

Cells are classified in two groups: primary and secondary. The grouping depends upon whether the cell can be recharged electrically or not.

A primary cell is one in which the energy available is obtained from chemical ingredients. When the chemical energy has all been converted into electrical energy the cell is exhausted. This type can be recharged only by renewal of the chemicals.

A secondary cell is 'reversible' in operation. If a current from a direct-current source is passed through the cell the plates undergo a chemical change and chemical energy is stored. This chemical energy can be reconverted as required into electric energy in an external circuit. The process of charging and discharging can be repeated indefinitely.

Primary cells

Leclanché cell (zinc–carbon)

The most common type of primary cell is the Leclanché. The e.m.f. (electromotive force) is about 1.5 V. Because of the high internal resistance and the reduction in voltage due to polarisation the cell is only suitable for supplying low-value intermittent currents. Typical applications are bell circuits and the speech circuits of telephones.

The dry cells are available in the small cylindrical type used in torches, and for bell circuits may be up to about 178 mm high by 76 mm diameter. The voltage is not affected by physical dimensions, but the larger cells supply higher values of current.

The dry cell is a manufactured article and

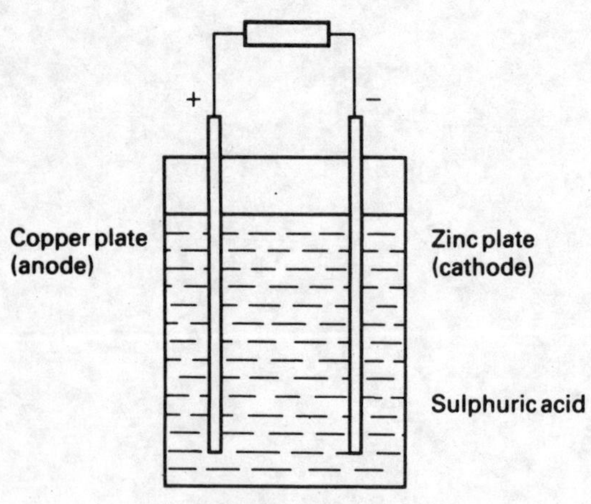

Figure 21.1. Simple cell.

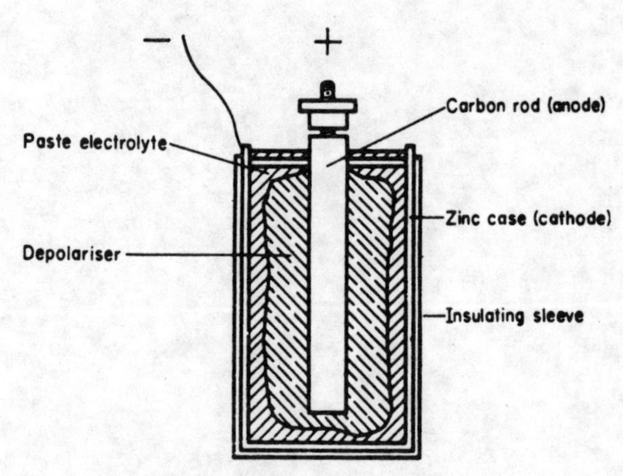

Figure 21.2. The dry Leclanché cell.

requires no maintenance. When exhausted the cell tends to swell and the paste electrolyte eats through the zinc casing. This paste attacks any metal near by and an exhausted cell should be discarded before any damage befalls equipment in which the cell is housed. A leakproof version is available in which a steel jacket covers most of the cell. This is slightly more expensive than the conventional dry cell, but protects equipment against the effects of corrosion.

When voltages greater than 1.5 V are required a battery is used consisting of the required number of cells connected in series. Small portable equipment, such as radios, use dry cells built up in layer-formation to give voltages ranging between 6 V and 90 V. The capacity of these layer-type batteries is small, being only about a hundred milliamps (100 mA) in the largest sizes.

Alkaline–manganese cells

This type of cell uses granulated zinc, high purity manganese dioxide and an alkaline electrolyte. The components are contained within a steel case, which does not take part in the chemical reaction and therefore results in a more effective cell seal. The cell can outlast the zinc–carbon cell by as much as six times in continuous use. Alkaline cells also have a storage life of about four years compared with that of the zinc–carbon, which is about a year or so.

The open-circuit (terminal voltage) output is 1.5 V with a discharge voltage of around 0.8 V. The alkaline battery excels in performance on continuous heavy loads and can operate efficiently between temperatures of between −30 °C and +70 °C. At zero temperatures the zinc–carbon cell virtually ceases to function.

Mercury cells

In order to supply appreciable currents the Leclanché cell must be made in large sizes. This becomes a problem in some cases due to the weight and the volume occupied by the cell. In cases where fairly high capacity cells are to be housed in a small space use is being increasingly made of mercury cells. These have a current rating of four times that of a dry Leclanché. The cell is available in several different outlines, but an important difference from the Leclanché is that the centre electrode is negative and the outer casing positive. The e.m.f. is about 1.35 V.

The mercury cell has many advantages, some of these being outlined as follows:

(*a*) Can be left unattended for very long periods. This makes it suitable for alarm systems and measuring instruments of other equipment which may stand idle for long periods.
(*b*) Does not suffer from polarisation and hence can supply current for a reasonable time.
(*c*) Virtually unaffected by temperature variation.
(*d*) Can withstand momentary overloads.

A possible disadvantage of the mercury cell is its initial cost. Since the life is very much greater than a Leclanché there is little difference, however, in the long-term cost.

In addition to the application given in (*a*) above, the small dimensions of the cell make it most suitable for use in small radios and deaf hearing-aids.

It must be remembered that the mercury cell is a primary cell and cannot be recharged electrically. Maintenance is practically nil, the cell being replaced when exhausted.

Secondary cells

Secondary cells, usually referred to as accumulators, in common use are the lead-acid and alkaline types. These cells have the great advantage that they can be recharged when exhausted from a suitable electricity supply. Their low resistance means they can supply large currents. Provided it is correctly maintained the life of an accumulator in regular use may be several years.

Lead-acid cells

The lead-acid cell consists of either a pair or two sets of lead plates immersed in a solution of dilute sulphuric acid. If current is passed through the cell from a direct-current source, such as a generator or even another battery, chemical changes occur at each plate. The plate connected to the supply positive becomes coated with lead peroxide. The negative becomes a type of spongy lead. Lead sulphate is removed from each plate and dissolves in the acid, thus increasing its strength (specific gravity). The conditions for a cell are now fulfilled since we have two different types of lead in an electrolyte. It will be found that if the cell is removed from the supply a voltage will exist between the terminals.

On discharge, that is if the cell is used to supply current to a circuit, the chemical energy stored in the cell is changed back into electrical energy. Both plates tend to turn into lead sulphate and the acid strength drops. In actual practice the accumulator is assumed to be completely discharged when the voltage falls to about 1.85 V. If the voltage falls below this the sulphate layer proves difficult to remove by charging and the accumulator is practically ruined.

Modern accumulators are usually of the multi-plate variety. The cell has an odd number of plates sandwiched together but separated from each other by wood or PVC separators. The positives are all joined together and connected to a common terminal. The negatives are connected in a similar manner (see Figure 21.3).

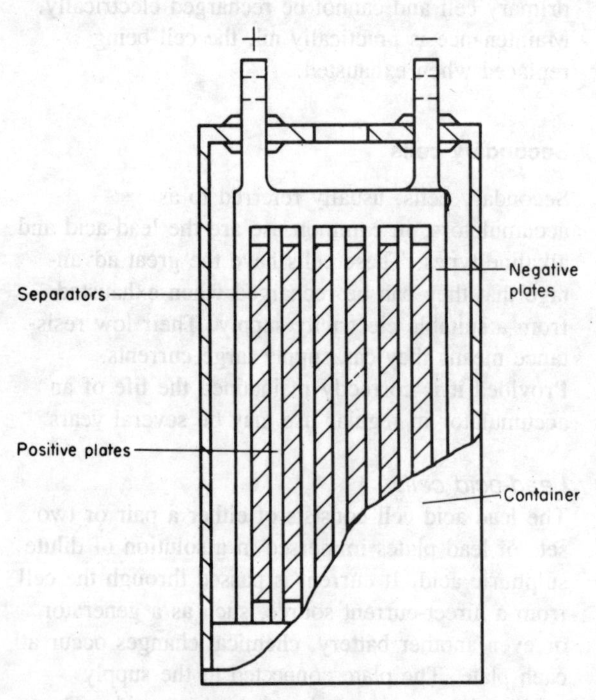

Figure 21.3. Lead-acid cell.

This type of cell has a high current capacity for its size because of the large surface area of the plates. The extra negative plate gives support to the positive plates which would otherwise tend to 'shed' their active material during use. Batteries used in motor vehicles are made up of multi-plate cells connected in series to give the required voltage. These large heavy-duty batteries have individual cells containing up to thirteen plates.

The voltage of a lead-acid cell when charged is about 2.2 V falling to a steady 2 V during use. As the cell approaches the end of the discharge period the voltage starts to fall rapidly to 1.85 V when it should be recharged. During use the plate colour changes; this change can be used as a guide to condition. When charged, the positive plate is a rich brown; the negative is slate-grey. In addition, bubbles can be seen rising from the plates (gassing). On discharge the plates both tend to become lighter in colour.

Applications of lead-acid cells include emergency service storage batteries such as used for stand-by lighting purposes. These large batteries may give voltages of up to 110 V by connecting a number of cells in series and are of heavy-duty rating.

The life of a lead-acid cell depends on efficient and regular maintenance and the principal points to observe are as follows:

(*a*) The currents on charge and discharge should not exceed those stated by the manufacturers.
(*b*) The cell should be recharged when the voltage approaches 1.85 V or when the specific gravity falls to about 1.18. The specific gravity is checked using a hydrometer and drawing a sample of acid from the cell.
(*c*) The level of the electrolyte must always be kept above the plates. If it is required to add liquid, to replace electrolyte lost by evaporation, only distilled water should be used. Acid of the correct strength may be added to replace any lost by spillage.
(*d*) Cells making up stand-by supplies (e.g. emergency lighting, switchboard protection supplies, etc.) should receive a trickle charge from a suitable dc supply.

(*e*) The tops of the cells should be kept clean and free from any liquid. The terminals should be lubricated with petroleum jelly to prevent corrosion.

(*f*) Any cell withdrawn from service and stored should be examined and given a charge at regular intervals. The current value for charging in this case is just a little over half that for normal use.

In stations where batteries are charged, provision should be made for adequate ventilation. Hydrogen gas is liberated during charging and no naked light must be used in any charging stations.

Alkaline cells

There are two type of alkaline cell in use; the nickel-iron and the nickel-cadmium. Both types are housed in steel cases and the electrolyte is caustic potash (potassium hydrate) and distilled water to give a specific gravity of about 1.2. The e.m.f. of these cells is about 1.4 V and the electrolyte strength does not change during operation.

In the nickel-cadmium cell the positive material is nickel hydroxide enclosed in pockets made from finely perforated steel ribbon. The negative material is cadmium oxide enclosed in similar steel-ribbon plates. This gives a flat plate formation. The nickel-iron cell differs in that the active materials are enclosed in steel tubes and iron oxide is used for the negative material.

Alkaline cells have the advantages of long life, saving in weight over the lead-acid cell, are virtually unaffected by momentary overloads and are very strong mechanically. They are, however, more expensive than acid cells although this is partly offset by their long life. Maintenance is similar to that of lead-acid cells, but not quite as critical. For example, an alkaline cell can discharge almost completely without ill effect, although this is naturally bad practice. One point to be borne in mind is that **the electrolyte is very corrosive** and can inflict serious damage if allowed to come into contact with the skin. Alkaline batteries are made up by connecting cells in series. The steel casings of the cells are electrically negative and batteries are constructed by housing each cell in a wooden or plastic insulating case.

Alkaline batteries can be used in practically any situation to replace acid batteries. Typical applications are batteries used in transport, emergency lighting systems and in miniature forms in rechargeable torches and transistors.

The mains points to be observed regarding storage batteries are summarised as follows:

(*a*) Reverse-current protection is to be provided where dc generators are connected in parallel with a battery. Effective means are to be provided to protect the batteries from excessive charging or discharging current. This protection can be a fuse or a circuit-breaker.

(*b*) Switchboards controlling the supply from a secondary battery installation must have a double-pole switch or circuit-breaker to isolate the battery from both load and charging circuits. If the charging circuit is used to share the load with the battery, means must be provided to isolate the charging circuit from the battery and the load. This can be a double-pole linked switch or circuit-breaker.

(*c*) Batteries should be arranged to be accessible from at least one side and the top. If the working voltage of the battery is 60 V or more, the individual cells must be supported on glass or porcelain insulators. The battery racks must be insulated if the working voltage is above 120 V. All connecting bolts should be lubricated with petroleum jelly unless of the non-corrosive type. Open-type cells to be fitted with spray-arresters to minimise the effects of acid spray particles in use.

(*d*) Celluloid shall not be used other than for portable batteries, and in the case of batteries using celluloid construction the charging arrangements shall be such that the effects of fire are minimised.

(*e*) Rooms containing secondary batteries shall be adequately ventilated to the outside air. If the batteries use sulphuric acid, the fittings must be non-corrosive or treated with acid-resisting paint.

It should be noted that reverse-current protection

is required when charging is from a dc generator. Should the driving motor or engine fail the batteries could send current in the 'reverse' direction through the generator causing it to 'motor', thus reversing the field polarity. Subsequent restoration of the supply would cause the generator to give out a voltage reverse to that previously obtained.

22 Lighting

Lighting plays a most important role in many buildings, not only for functional purposes (simply supplying light) but to enhance the environment and surroundings. Modern offices, shops, factories, shopping malls, department stores, main roads, football stadia, swimming pools — all these show not only the imagination of architects and lighting engineers but the skills of the practising electrician in the installation of luminaires.

Many sources of light are available today with continual improvements in lighting efficiency and colour of light. Two terms associated with lighting will be used in this chapter:

lumen (lm): this is a unit of luminous flux (or 'amount of light') emitted from a source.
luminous efficacy: this denotes the amount of light produced by a source for the energy used; therefore the luminous efficacy is stated in 'lumens per watt' (lm/W).

A number of types of lamps are used today: filament, fluorescent, mercury vapour, sodium vapour, metal halide, neon. All these have specific advantages and applications.

Filament lamps

These use a filament made from tungsten and raised to about 2,500 °C to produce a light which, though it looks white, actually has a lot of red in it. Various methods are used to produce an efficient light-emitting filament (coiled, coiled-coil, etc.). The common filament lamps used today have a luminous efficacy of about 16 lm/W. Slightly higher outputs are available when the light bulb is filled with an inert gas such as argon. One problem with filament lamps is the tendency of the tungsten to evaporate; the result of this process is seen in the blackening of the inside of the glass bulb. Another problem is the heat produced by the lamp. In general filament lamps use much more energy than other types of lamp, such as the fluorescent.

The nominal life of a filament lamp is 1,000 hours. But this figure is influenced by a number of factors. For instance, a 240 V lamp operating at 250 V will reduce the life expectancy by 43 per cent. Frequent switching also tends to reduce the life of a lamp, as will vibration. Because of the relatively low life expectancy of a filament lamp, lamp replacement costs tend to be higher than those for, say, fluorescent lamps (7,500–10,000 hours burning).

The standard type of filament lamp is the GLS (general lighting service). Two cap fittings are available: BC (bayonet cap) and ES (Edison screw). The former type of fitting is used for lamps rated from 15 W to 150 W; above this size the ES fitting is used.

Incandescent lamps are used for many purposes and are available with many variations. 'Pearl' lamps have the glass bulb internally frosted. Other types have the glass bulb silica-coated internally. The following are some of the applications of filament lamps:

General lighting. Vacuum or gas-filled lamps; top-reflector lamps (frosted and half-mirrored inside).

Festive lighting. Spherical, candle and chandelier lamps, with frosted or opalised finish; coloured and white.

Spot and flood lighting. Lamps made from pressed glass and internally mirrored to radiate a definite beam of light. The floodlight has a relatively broad beam. The spotlight has a narrow beam. These lamps are very strong and are used for shop and showcase illumination. Floods are used for outdoor illumination (gardens, parks and sports grounds).

Reinforced construction lamps. These lamps are intended for use where vibration and shock are more than normal: usually called rough-service lamps.

Signal lamps. These are used for signalling purposes on switchboards and for indication installations: sometimes called 'pygmy' lamps.

Thermal radiation lamps. These lamps are used in piglet- and chicken-rearing. Heating lamps are hard-glass bulbs internally mirrored for use for short periods at a time. They are used for heating in bathrooms and in industry for drying processes (e.g. stove-enamelling).

Special lamps. In this group are included sewing-machine lamps, oven lamps, showcase lamps and Christmas-tree illumination sets.

One type of lamp of recent introduction is the quartz-iodine lamp, in which iodine vapour is used to control the rate of evaporation of the filament material and thus prolonging its life. This type of lamp is smaller than other types of filament lamps,

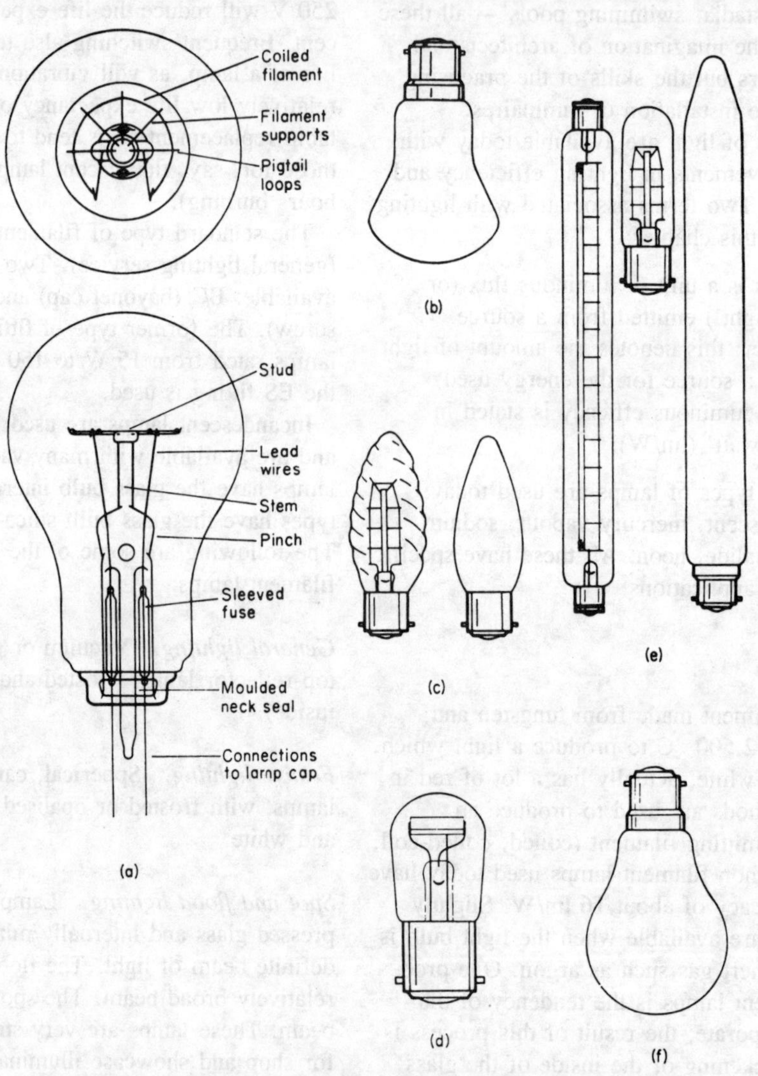

Figure 22.1. (a) The filament lamp and its construction; (b) The 'mushroom' filament lamp; (c) Candle filament lamps for decorative lighting; (d) Neon lamp; (e) Various types of tubular filament lamps; (f) Modern filament lamp.

though the problem of heat dissipation is increased. Usually the metal housing of the fitting is of finned construction, and the terminal chamber for cable entry is partially separated from the main housing. The main application of this lamp is for floodlighting. The reflectors must be protected by toughened glass, because there can be a considerable difference in temperature between the edge and the centre.

Discharge lamps

Under normal circumstances, an electric current cannot flow through a gas. However, if electrodes are fused into the ends of a glass tube, and the tube is slowly pumped free of air, current does pass through at a certain low pressure. A faint red luminous column can be seen in the tube, proceeding from the positive electrode; at the negative electrode a weak glow is also just visible. Very little visible radiation is obtainable. But when the tube is filled with certain gases, definite luminous effects can be obtained. One important aspect of the gas discharge is the 'negative resistance characteristic'. This means that when the temperature of the material (in this case the gas) rises, its resistance decreases — which is the opposite to what occurs with an 'ohmic' resistance material such as copper. When a current passes through the gas, the temperature increases and its resistance decreases. This decrease in resistance causes a rise in the current strength which, if not limited or controlled in some way, will eventually cause a short circuit to take place. Thus, for all gas-discharge lamps there is always a resistor, choke coil (or inductor) or leak transformer for limiting the circuit current. Though the gas-discharge lamp was known in the early days of electrical engineering, it was not until the 1930s that this type of lamp came onto the market in commercial quantities. There are two main types of electric discharge lamp:

(*a*) Cold cathode.
(*b*) Hot cathode.

The cold-cathode lamp uses a high voltage (about 3.5 kV) for its operation. For general lighting purposes they are familiar as fluorescent tubes about 25 mm in diameter, either straight, curved or bent to take a certain form. The power consumption is generally about 8 W per 30 cm; the current taken is in milliamps. The electrodes of these lamps are not preheated. A more familiar type of cold-cathode lamp is the neon lamp used

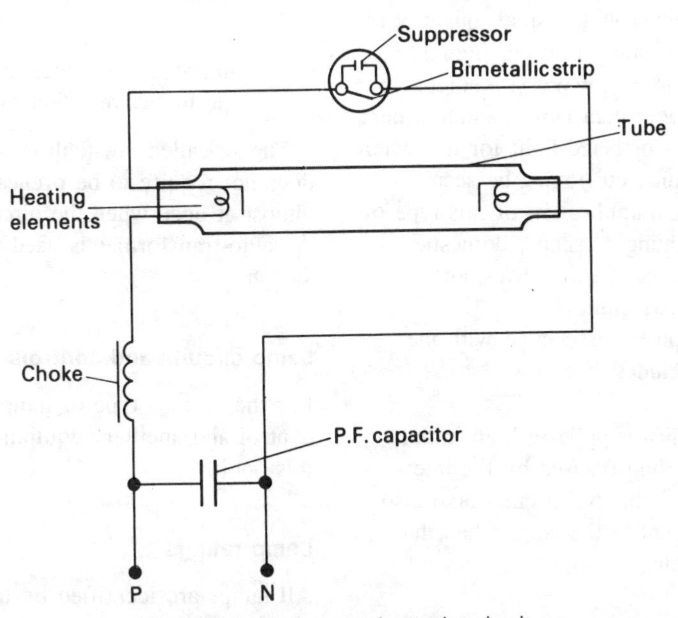

Figure 22.2. Fluorescent-tube starting circuit.

for sign and display lighting. Here the gas is neon which gives a reddish light when the electric discharge takes place in the tubes. Neon lamps are also available in very small sizes in the form of 'pygmy' lamps and as indicating lights on wiring accessories (switches and socket-outlets). This type of lamp operates on mains voltage. Neon signs operate on the high voltage produced by transformers.

The hot-cathode lamp is more common. In it, the electrodes are heated and it operates generally on a low or medium voltage. Some types of lamp have an auxiliary electrode for starting. Other types, more within the scope of this chapter, called 'switchstart' lamps, use a switching arrangement in the circuit (see Chapter 20).

The most familiar type of discharge lamp is the fluorescent lamp. It consists of a glass tube filled with mercury vapour at a low pressure. The electrodes are located at the ends of the tube. When the lamp is switched on, an arc-discharge excites a barely visible radiation, the greater part of which consists of ultra-violet radiation. The interior wall of the tube is coated with a fluorescent powder which transforms the ultra-violet rays into visible radiation or light. The type of light (that is, the colour range) is determined by the composition of the fluorescent powder. To assist starting, the mercury vapour is mixed with a small quantity of argon gas. The light produced by the fluorescent lamp varies from 45 to 55 lm/W. The colours available from the fluorescent lamp include a near-daylight and a colour-corrected light for use where colours (of wool, paints, etc.) must be seen correctly. The practical application of this type of lamp includes the lighting of shops, domestic premises, factories, streets, ships, transport (buses), tunnels and coal-mines.

The auxiliary equipment associated with the fluorescent circuit includes:

(*a*) the choke, which supplies a high initial voltage on starting (caused by the interruption of the inductive circuit), and also limits the current in the lamp when the lamp is operating;

(*b*) the starter;

(*c*) the capacitor, which is fitted to correct or improve the power factor by neutralising the inductive effect of the choke.

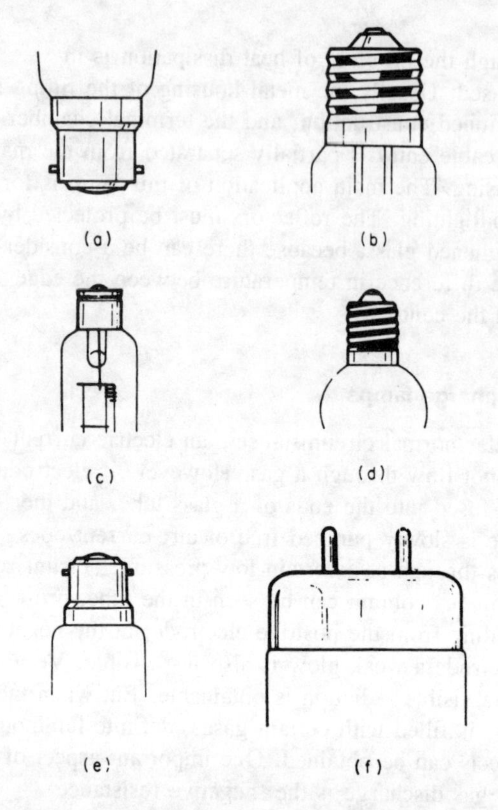

Figure 22.3. Types of lamp caps: (a) Spigot-type bayonet-cap; (b) Large Edison Screw; (c) non-spigot centre-contact; (d) Small Edison Screw, centre contact; (e) Small bayonet-cap, spigot-type; (f) Bi-pin type (fluorescent tube).

The so-called 'switchless' start fluorescent lamp does not require to be preheated. The lamp lights almost at once when the circuit switch is closed. An auto-transformer is used instead of a starting switch.

Lamp circuits and controls

For the various type of lamp circuits, and their control and ancillary equipment see Chapter 27 of this book.

Lamp ratings

All lamps are identified by their 'wattage', or the quantity of electrical energy they take to produce a definite quantity of light. Note, however, that carbon lamps are rated in 'candle-power'. This term is a flashback to the days when the standard light-producing device was a candle made from spermicetti wax. Nevertheless, filament lamps are rated at so many watts at a specified or rated voltage. The light output of lamps is very dependent on the supply voltage. Taking a light output of 14 lm/W, a 60 W, 240 V lamp will produce 14 × 60 lm = 840 lm. The current taken is $I = W/V = 60/240 = 0.25$ A. At this current we can calculate the resistance of the filament, $R_f = V/I = 240/0.25 = 960$ ohms.

Suppose now the supply voltage was increased to 260 V. The current would rise to $I = V/R_f = 260/960 = 0.27$ A, and the wattage taken to $W = VI = 260 \times 0.27 = 70$ W. The light output would also increase to 14 × 70 = 980 lm. The opposite would, of course, occur if the supply voltage were to fall below the rated voltage. Briefly, then, the light output depends on the voltage of the supply. All filament lamps have a nominal life of 1,000 hours, that is, each lamp will burn at rated voltage for an average of 1,000 hours before failing. To 'overvolt' a lamp (by increasing the supply voltage or by connecting, say, a 240 V lamp to a 260 V supply), will increase its light output, but reduce its useful life. Even overvolting by 5 per cent (e.g. a 230 V lamp on 242 V) will halve its life. Undervolting a lamp will reduce its light output and prolong its life; but this does not result in a corresponding reduction in the wattage consumed and electricity is in effect being wasted.

The nominal life of a fluorescent lamp is between 5,000 hours and 7,500 hours. With switched or switchless starting gear, the control gear losses generally amount to about 15 per cent of lamp watts. Thus, the circuit of an 80 W fluorescent lamp will take from the supply a total of 92 W.

Lamp summary

Filament: 16 lm/W; 1,000 hours nominal burning; lamp caps: BC or ES; colour: white with red content; types available: general ligting service (GLS); rough service (RS); pygmy (15 W rating); fireglow (for fire-effect electric heaters); coloured (for outside decorations); architectural (linear lamps used for mirrors, display cabinets, etc.); reflector lamps.

Halogen: 25 lm/W; 2,000 hours nominal life; colour: white with reddish content; wattages from 200 W to 2 kW; larger rtaings used for floodlighting.

Fluorescent: 75 lm/W; up to 10,000 hours nominal life; colours can include 'warm white', 'daylight' and 'northlight' (used where colour rendering is important, such as in textiles, painting, printing; cap: bi-pin).

Compact fluorescent: These are small single-ended fluorescent lamps which are generally replacing the GLS lamp; 8,000 hours nominal life; 60 lm/W; lamp caps: both BC and ES.

Light measurement

The amount of light falling on a surface is measured (in lumens per m^2; unit = lux) by an instrument called a photometer or light meter. It consists of a cell made from three layers of metal: (*a*) a transparent film of gold; (*b*) a film of selenium; (*c*) a steel baseplate.

A connecting ring makes contact with the transparent film. Another connection is taken from the steel plate. These connections are taken to a very sensitive moving-coil instrument, which has a scale graduated in lm/m^2. When rays of light fall on the surface of the cell between the gold and selenium films, electrons are freed, to cause a current to flow in the moving-coil of the meter movement. This current is approximately proportional to the amount of light falling on the cell. The instrument is used to check that the amount of light falling on working planes (tables, desks, benches) is sufficient for a particular job to be done with no strain on the eyes of the worker. The Illuminating Engineering Society publishes tables indicating the optimum amount of light required to perform various tasks in industry and in the home.

Regulations summary

If any circuit is used predominantly for discharge lighting, the neutral conductor must be of the same size as the phase conductor. Mineral-insulated cables should not be used for discharge lighting circuits unless surge arresters are used. The reason for this is that the inductive currents associated with such circuits produce high voltage surges which could puncture the mineral insulation and earth the sheath. When ES lampholders are used, the centre contact must always be connected to the phase conductor.

When bulkhead or well-glass luminaires are installed on the exterior wall of a building, they are regarded as being outside the equipotential zone (the earthing and bonding of all metalwork in the building). Circuits supplying such luminaires must have a disconnection time of 0.4 second if an earth fault occurs. If the luminaire is taken off a circuit within the building, the disconnection time is also 0.4 second.

All insulated conductors taken into a bulkhead luminaire must be sheathed with heat-resisting sleeving (this is because of the significant heat build-up inside the fitting).

When calculating the current taken by fluorescent luminaires, account must be taken of the fact that the associated circuits take more current than is indicated by the lamps' wattage, thus

$$I = \frac{\text{total wattage} \times 1.8}{\text{voltage}} \text{ amps}$$

where the factor of 1.8 includes such factors as the power factor of the circuit and the wattage loss in the choke (inductor).

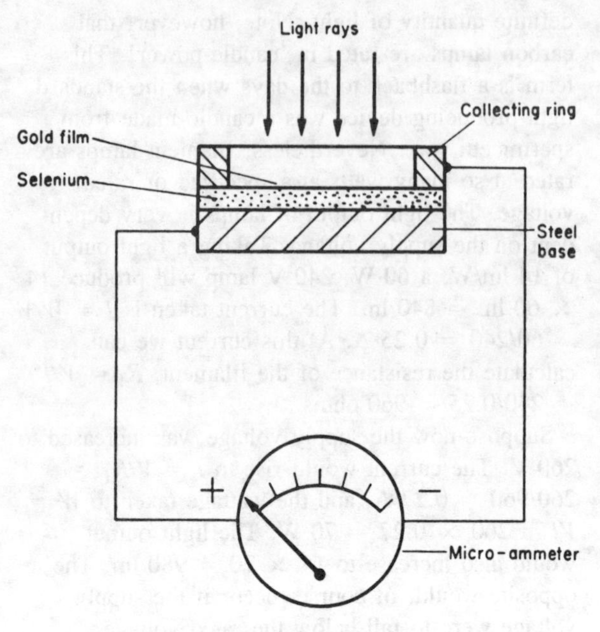

Figure 22.4. The light meter.

Switches for fluorescent luminaire circuits should be rated for inductive loads; if they are not, their rating must be reduced by a factor of 0.5.

All BC lampholders must have the correct temperature rating 'T2' which allows for a temperature more than 165 °C.

23 Electromagnetic devices

In electrical engineering work, extensive use is made of various types of electromagnetic devices for providing a mechanical force to operate different kinds of mechanism. Operation by electromagnetic means gives the advantage of automatic control with possible remote control. There are three main types of electromagnet:

1. *Tractive type*. In this type, the electromagnet attracts an armature to which the load mechanism is attached. An example is the contactor or relay, where the attraction and release of the armature activates one or more sets of contacts.

2. *Solenoid type*. In this type, the operating coil surrounds a sliding core or plunger which is drawn into the coil, usually in a vertical direction, when the coil is energised.

3. *Lifting type*. In this type, the poles of the electromagnet are brought into contact with magnetic material so that the material can be transported.

In addition to these types, there are the applications in which the electromagnet forms an integral part of a complete mechanism, such as a magnetic brake or clutch. The underlying principle is the same in all cases — based on the laws of electromagnetism applying to the attraction and repulsion of magnetised surfaces.

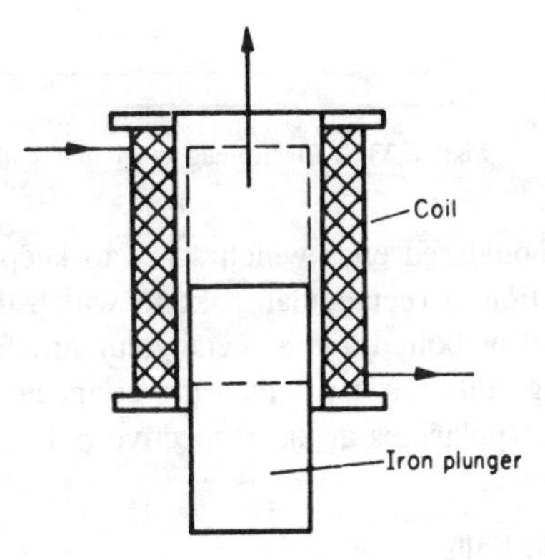

Figure 23.1. Solenoid and plunger.

Solenoids

The simplest form of electromagnet solenoid is a cylindrical coil, within which the bore of an iron plunger is free to move. The pull extended on the plunger with this simple arrangement when the coil is energised is very small. The pull is increased by making the coil long and increasing the cross-sectional area of the plunger. Another method of increasing the pull of the solenoid is to provide the solenoid with an iron circuit to complete the magnetic path as far as possible with the exception of a working air-gap. This iron circuit is generally a frame bent to shape from steel strip. In this form, the solenoid has a wide application for overcurrent and low-voltage relays. In an ac solenoid, the flux provided by the coil passes through a maximum and falls to zero twice in each cycle, so that the pull is not constant. Thus, an ac solenoid is not so efficient as a dc type.

Electromagnets

The general form of the electromagnet is shown in Figures 23.2 and 23.3. There is a cylindrical pole

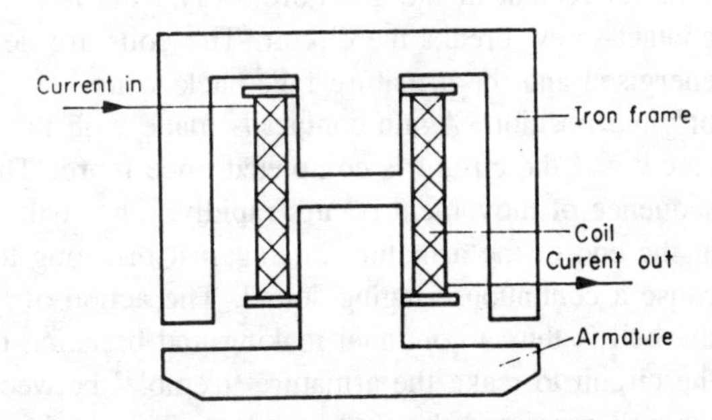

Figure 23.2. Electromagnet, single coil.

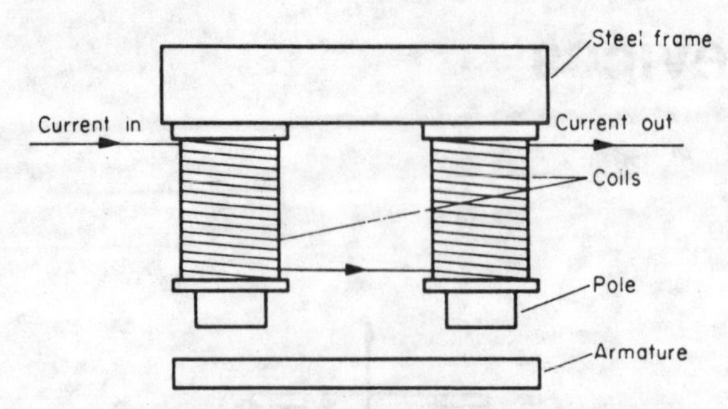

Figure 23.3. Electromagnet, double coil.

with shouldered ends which serve to keep the coil in position, a rectangular yoke to which the pole is screwed or bolted and a rectangular armature. Two exciting coils are used; they are connected to give opposite polarities at the respective pole ends.

Electric bells

There are two types of electric bell in common use:

1. Trembler bell
2. Continuous-ringing bell

1. *Trembler bell.* See Figure 23.4. This type of bell is in very common use as a door bell. It is similar in construction to the single-stroke bell except that a make-and-break arrangement is provided in the circuit. Instead of the coils being connected directly across the supply through the bell-push, the coils are in series with an adjustable contact-screw against which an armature spring-leaf contact normally rests. When the push is pressed, the current flows through the adjustable contact-screw, the armature and the coils. The coils become energised and attract the armature. This movement of the armature away from the contact-screw breaks the circuit. The coils are de-energised and the armature falls back to its original position. Again contact is made with the screw and the circuit is completed once more. This sequence of movement occurs rapidly. The striker at the end of the armature hits against the gong to cause a continuous ringing sound. The action of the bell is thus a continual making and breaking of the circuit to make the armature 'tremble' between the coil cores and the contact-screw. The ringing

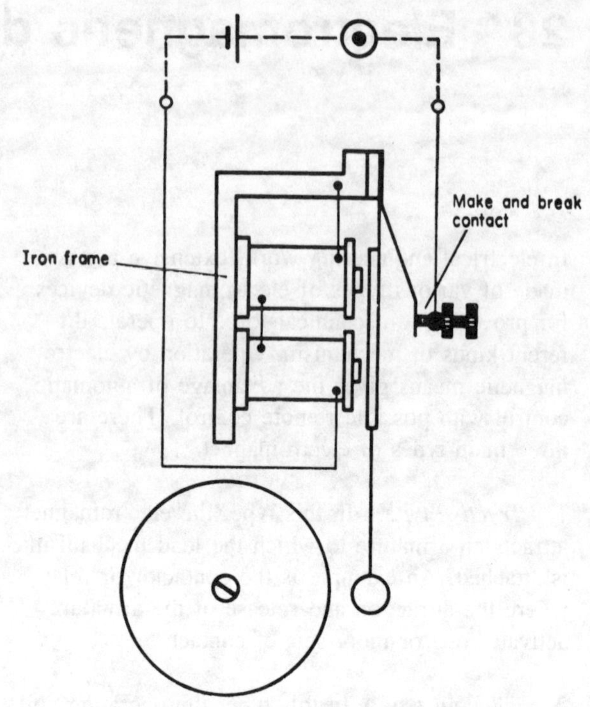

Figure 23.4. Trembler electric bell.

will continue for as long as the push is pressed. Because sparking occurs at the contact points, these are often made of silver or platinum, which do not oxidise easily. The tone of the sound produced by the bell can be changed by altering the setting of the adjustable contact-screw.

2. *Continuous-ringing bell.* See Figure 23.5. This type is basically a trembler bell, but with either a mechanical or electrical arrangement to make the bell continue to ring after the bell-push has been released. A small lever is placed below the contact-screw. On the first movement or stroke of the armature, this lever drops automatically and, as it drops, it short-circuits the bell-push, thus causing the bell to ring continuously. The lever can be reset by pulling a cord, so stopping the bell.

Indicators

Often a bell circuit is arranged so that the bell can be operated from any one of two or more positions. In such circuits it becomes necessary to install an indicator board to show which bell-push

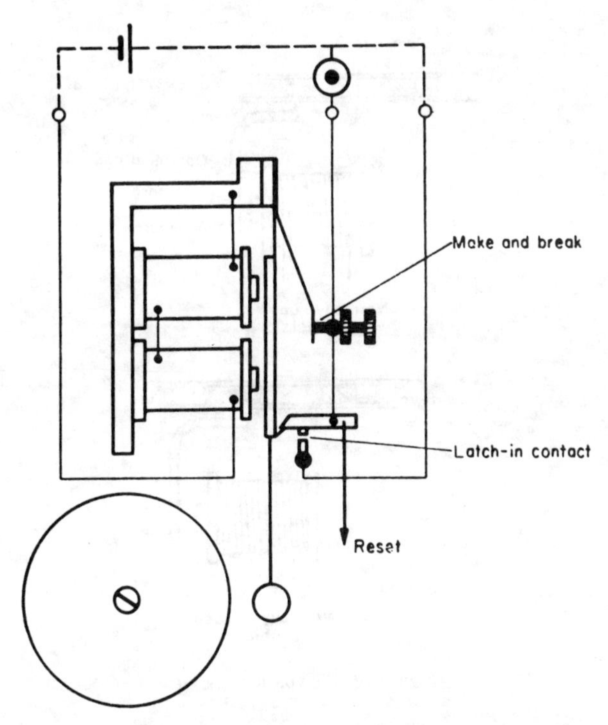

Figure 23.5. Continuous-ringing electric bell.

has been operated. There are three main types of indicator:

1. *Pendulum type*. The movement of this type is similar to that of the single-stroke bell. A soft-iron armature is pivoted or hinged at one end; the other end carries a flag. The armature is located in front of the electromagnet, the coil of which is connected in series with a bell-push. When the push is pressed, the electromagnet becomes energised and it attracts the armature. As the adjustable contact-

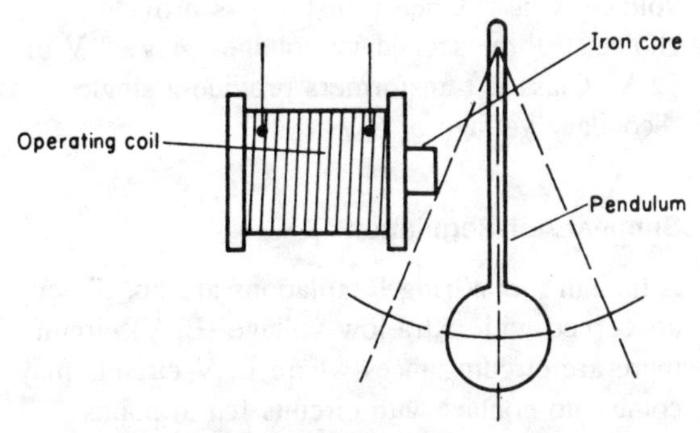

Figure 23.6. Pendulum indicator unit.

screw on the associated bell makes and breaks the circuit, so does the electromagnet become energised and de-energised. And the armature swings to and fro in pendulum fashion for a short time before coming to rest. The movement of the flag is seen through a clear circular part of the glass screen. The disadvantage of this type of indicator is that if the person called happens to be out of the room when the bell rings, the pendulum may well have stopped swinging by the time the person returns and reaches the indicator.

2. *Mechanical-replacement type*. In order that some indication that a bell has rung may be displayed for an indefinite time on the indicator board, it is necessary to have some arrangement for the manual replacement of the indicator flag after the request for attention has been satisfied. One method used is by mechanical means.

The mechanical-replacement type of indicator element has a single-core electromagnet. The armature is pivoted near its centre and held away from the magnet core by a spring at the end remote from the magnet. The flag is attached to one end of a pivoted arm which has an extended piece caught normally by a catch on the armature. When the bell-push is pressed, the electromagnet is energised and attracts the armature. The movement of the armature causes the catch on the pivoted arm to be released and the flag moves to show itself in the appropriate space in the window of the indicator board. The flag will remain in this position even though the bell-push is released. To reset the flag, a lever is either pushed or turned round by hand.

3. *Electrical-replacement type*. This type of movement has two separate electromagnets. One coil is in series with the appropriate call bell-push as before. The other coil is connected to a 'replacement' circuit with a separate bell-push. The armature, to which the flag arm is directly attached, is pivoted at its centre in such a way that it can be attracted by either magnet. When the call push is pressed, the armature is attracted to the alarm or indication position. The flag shows in the window of the indicator board. When the replacement push is operated, the armature is attracted by

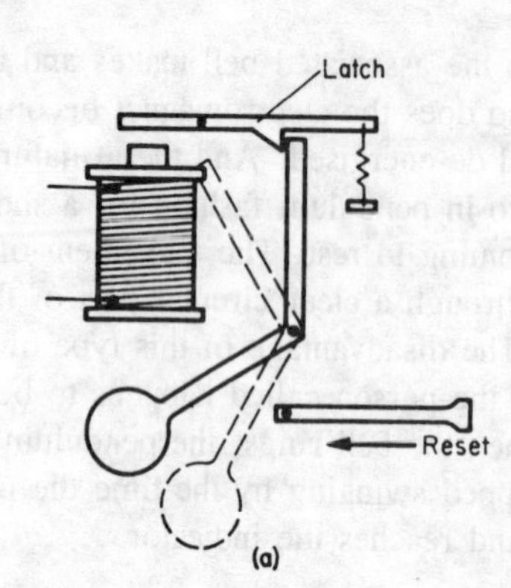

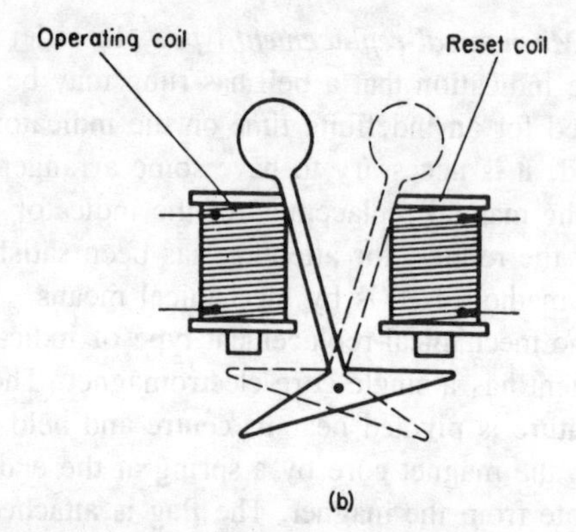

Figure 23.7. (a) Flag indicator unit, mechanical reset; (b) Flag indicator unit, electrical reset.

the other magnet and the movement causes the flag to move out of sight, to its original position.

Bell relays

When a bell circuit is long in length, it may be found that the volt drop along the circuit wires is so great that the bell will not operate, or else does so inefficiently. Additional cells or batteries may be used to raise the circuit voltage, to compensate for the lost volts. But this may not always be a satisfactory solution to the problem. The other solution is to employ a relay. The relay is basically a pair of electrical contacts operated by an electromagnet. A very small current is sufficient to energise the electromagnet and attract a spring-controlled armature. The movement of the armature causes the two contacts to close and operate the bell. The relay circuit may be energised separately from its own supply, or it may use the bell supply. Operation of the bell-push will energise the relay and cause the bell to ring. Typical relay circuits are shown in Chapter 27 of this book.

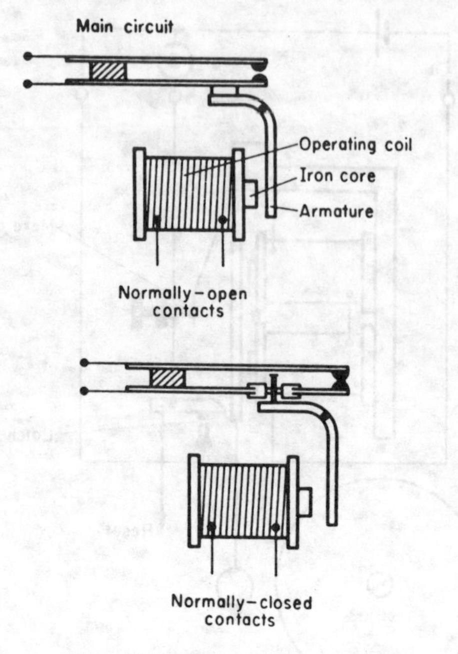

Figure 23.8. Typical relay types.

Bell transformers

The simplest bell circuit uses a cell or battery for its source of energy. The bell may also be supplied from a double-wound transformer. It consists of a soft-iron core with two windings: the primary and the secondary. The primary winding accepts the mains ac voltage. The secondary winding, which has no electrical connection with the primary, produces (by mutual induction) a low voltage. Class A bell transformers provide a choice of three secondary voltages: 4 V, 8 V or 12 V. Class B transformers provide a single secondary voltage of 6 V.

Summary of Regulations

Although the Wiring Regulations are not directly concerned with extra-low voltage (ELV) circuits, there are circumstances where ELV circuits may come into contact with circuits fed at mains voltage.

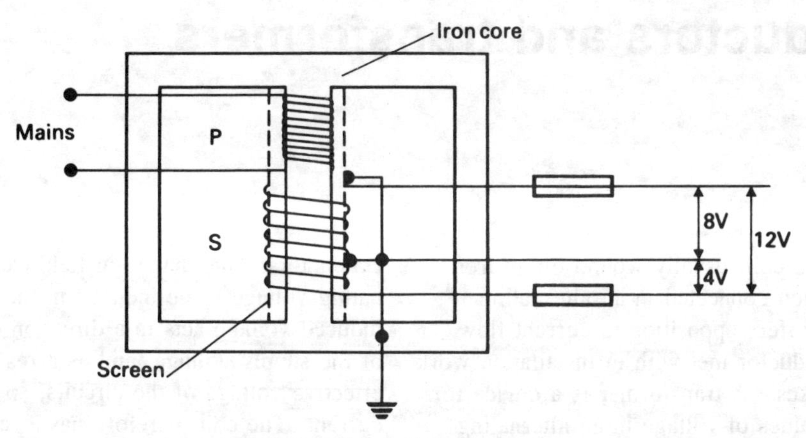

Figure 23.9. Bell transformer.

In electrical installations, three categories of circuit are recognised:

Category 1: Those fed from mains voltage, other than fire alarm and emergency lighting circuits;
Category 2: ELV circuits such as radio, telephone, bell and call circuits;
Category 3: Fire alarm and emergency lighting circuits.

It is required that these three categories of circuits must always be segregated, that is they may not be taken into a common enclosure such as conduit or uncompartmented trunking. Category 1 and 2 circuits may, however, be run in the same enclosure provided that the Category 2 (ELV) circuits are insulated for mains voltage.

All bell transformers must have provision for the enclosure of the cores of the supply cable and terminals must be provided with a cover to prevent terminals being touched. Some bell transformers are regarded as a functional ELV circuit in which case one point of the secondary winding and the exposed conducting parts (the transformer core) must be connected to the protective conductor (CPC) of the mains circuit feeding the transformer. If, however, the transformer is double-insulated, carrying the British Standard symbol for double insulation (two squares, one inside the other), the earth connection can be omitted.

Most bell transformers tend to be connected into the lighting circuit because their rating is usually not greater than 5 VA (5 volt-amperes). Otherwise they should be taken from a separate circuit in the consumer unit.

24 Inductors and transformers

An inductor is a coil, usually wound on an iron core, which when connected in an alternating current circuit offers opposition to current flow. The types of inductor met with in installation work are called 'chokes'. A transformer is a device for changing the values of voltage in an alternating current circuit. Both pieces of apparatus depend for their operation on induced voltages set up by a fluctuating magnetic field due to the ac supply system. A brief indication of the principles involved is given below.

Self-induction

When a conductor, or coil, is moved in a magnetic field, an e.m.f. is set up in the conductor. The e.m.f. is only present as long as the conductor is moving. Similarly, the conductor can be stationary and the magnetic field altered, giving the same effect. On any circuit carrying current, a magnetic field is set up due to the current. Figure 24.1 shows a coil carrying current and the resulting magnetic field.

The magnetic field alternates with the same frequency as the current and embraces the coil-conductors. The change in field causes an alternating voltage to be induced in the coil. This induced voltage acts in a direction opposite to that of the supply voltage, and as a result reduces the effective voltage of the circuits, so limiting the current. The coil therefore has a 'choking' effect, and offers an opposition to current flow called the inductive reactance.

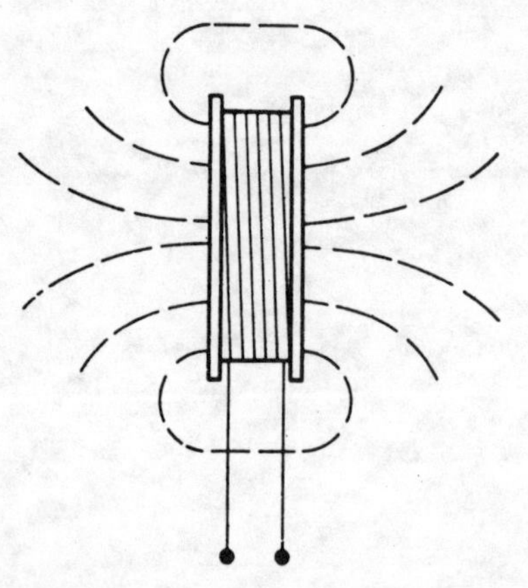

Figure 24.1. Current-carrying coil and its magnetic field.

Any circuit in which a change in current produces an induced e.m.f. possesses 'inductance' and the circuit is an 'inductive' one. The voltage induced in this case is due to changes in a single circuit and is referred to as being 'self-induced'.

When an inductive circuit is used on direct-current and is switched on or off, the sudden change in field strength causes a self-induced e.m.f. to be generated. The induced e.m.f. can be greater than the applied voltage depending upon how great is the rate of change of field. This induced voltage can show itself as arcing at the contacts of switches. Direct-current tumbler switches have a faster action and wider contact clearance than ac switches of equal rating to minimise effects of arcing.

Applications

The problem of arcing is not so critical on ac, but voltage surges are still produced on switching. This effect is put to use in the starting of fluorescent discharge lamp circuits. A choke is connected in series with the lamp and operation of the starting switch causes a momentary high voltage to be induced. The high voltage initiates the discharge and allows the lamp current to flow. Once the lamp strikes, the choke then limits the circuit current to the value required to operate the lamp. Discharge lamp circuits and a description of these lamps are given in Chapter 22.

Series chokes, or inductors, are also used in dimmer circuits to reduce the intensity of illumination in stage lighting, etc. The resistance of a choke is small compared to its inductive reactance when used on ac. The reduction in voltage across the lamps is achieved with the power loss and heating which occurs in a resistor dimming circuit. It must be kept in mind, however, that the power factor of the circuit is reduced by the use of the highly inductive choke.

The requirements of the Regulations applying to the installation and operation of chokes are given at the end of this chapter.

Mutual induction

Consider two coils, wound on a common iron core and one of them connected to the supply. If the switch is closed a magnetic field links both coils as shown in Figure 24.2. The opening or closing of the switch will cause a self-induced e.m.f. to be generated in the coil connected to the supply. Since the field also links the other coil, an e.m.f. is induced in that coil by the changing field. There is no electrical connection between the two coils and the second e.m.f. is said to be 'mutually induced'.

Where a change of current in one circuit results in an e.m.f. being induced in another circuit, the two circuits are said to possess 'mutual-inductance'.

If the direct current source is replaced by an alternating voltage, an alternating field is set up.

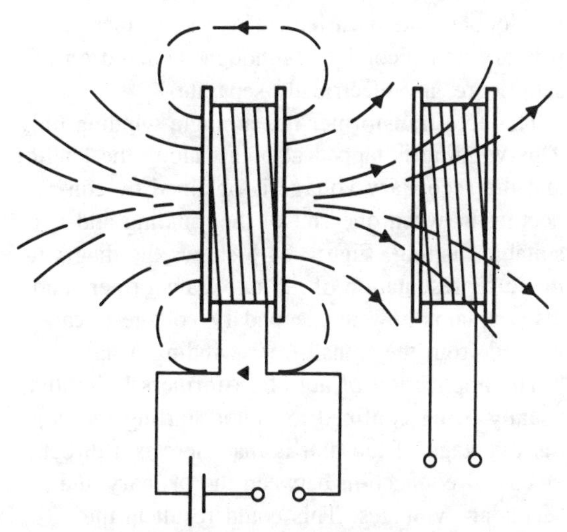

Figure 24.2. Mutual induction between two coils.

This causes an alternating voltage to be induced in the second coil having the same frequency as the supply. The double-wound transformer operates on this principle. The coil connected to the supply is the primary, and that in which an e.m.f. is mutually induced is the secondary.

Transformers can be used to decrease or increase the supply voltage. Transmission is usually high voltage and this is reduced to working voltages by step-down transformers. The generated voltage of a power station is increased in the opposite manner using a step-up transformer. The amount of increase or decrease depends upon the number of turns on the primary and secondary coils. For example, if the primary has 2,000 turns and the secondary 1,000 turns the secondary voltage will be half the applied voltage.

It must be noted that the transformer operates on ac only. For an e.m.f. to be continually induced in the secondary the field must be continually varying. This is possible if the supply voltage is alternating.

Transformer construction

The coils of the transformer are wound on a closed iron core. To minimise the losses in the core due to the alternating current, it is made up of laminations (thin steel strips lightly insulated from each other). The basic diagram of a transformer is shown in Figure 24.3 in which each coil occupies one side of a rectangular core. In practice, the coils are either wound over each other (concentric) or wound in sandwich formation. Two types of core formation are commonly used: the core type using a single magnetic circuit or the

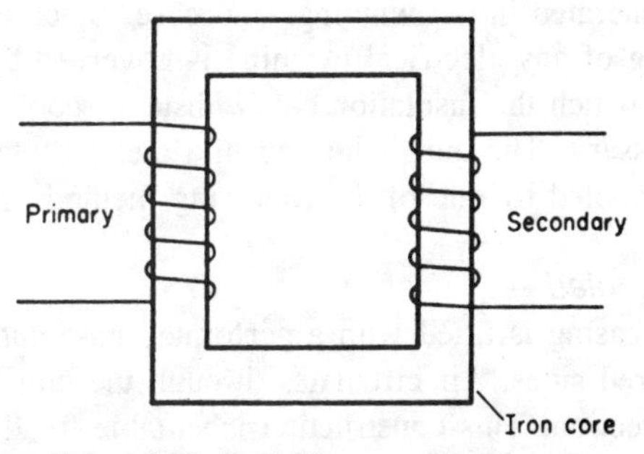

Figure 24.3. Basic diagram of transformer.

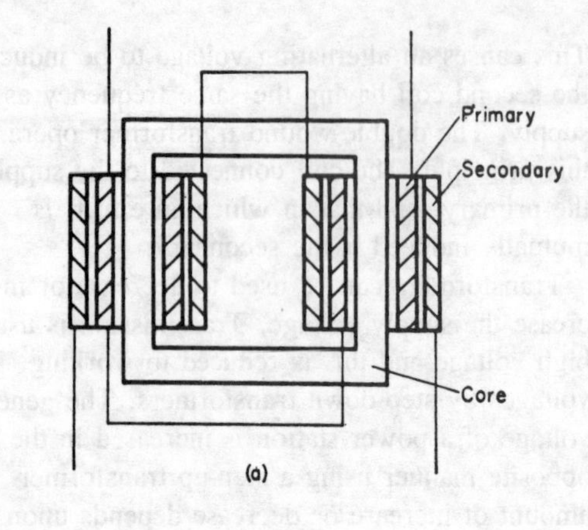

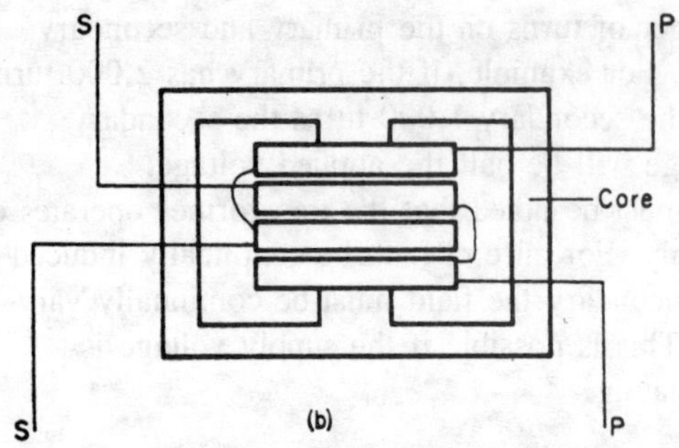

Figure 24.4. (a) Core-type transformer; (b) Shell-type transformer.

shell type with a double magnetic circuit. Winding and core arrangements are shown in Figure 24.4.

In order to compensate for volt drop in cables and variations in supply voltage the secondary is usually tapped. These tappings are arranged to give voltages above and below the nominal secondary voltages. Typical values of tapping give $\pm 2\frac{1}{2}$ per cent and ± 5 per cent of the nominal voltage. The tappings are connected via removable links on the secondary.

When the transformer is supplying a load, heat is generated in the windings and core. Since the rating of any electrical machine is governed by the heat which the insulation can withstand, cooling is necessary. The unit is housed in a metal container and cooled by one of the following methods:

Air-cooled

The casing is fitted with a perforated base and louvred sides. Air circulates through the unit by convection. This construction is suitable for dry dust-free conditions and is usually confined to small transformers of about 5 kVA (kilo volt amp) or less.

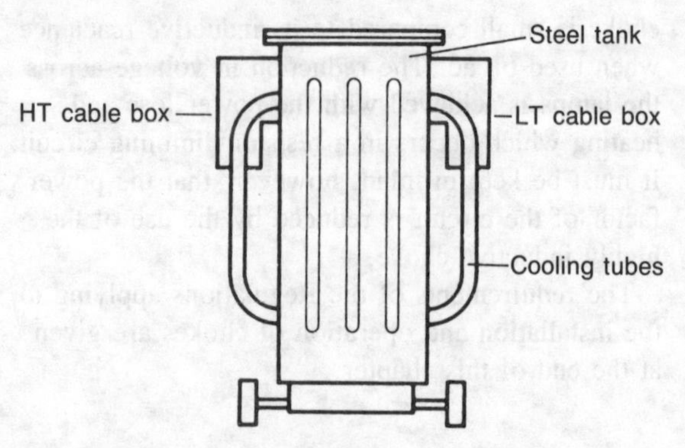

Figure 24.5. Typical oil-filled transformer.

Oil-filled

The transformer is housed in a casing containing insulating oil which completely covers core and windings. The oil acts as a cooling and insulating medium. Pipes are fitted around the casing and the air passing over these pipes carries away the heat that has been transferred from the core through the oil. Figure 24.5 gives the construction of a typical oil-filled transformer. This is the most common construction.

Auto-transformers

The transformers previously described are all of the double-wound variety. This means that the primary and secondary, although mounted on the same core, are electrically separate.

The auto-transformer has a single winding only. This winding is tapped at points along the length, and the secondary voltage is obtained by connecting between one end of the winding and a suitable tapping. Figure 24.6 shows the diagrammatic representation of an auto-transformer, and it also explains how the secondary voltage is calculated from the transformer winding turns.

The application of auto-transformers is limited, usually being confined to motor starting. A major disadvantage of the unit is that there is a direct electrical connection between the primary and secondary voltages. This could result in the primary voltage appearing at the secondary ter-

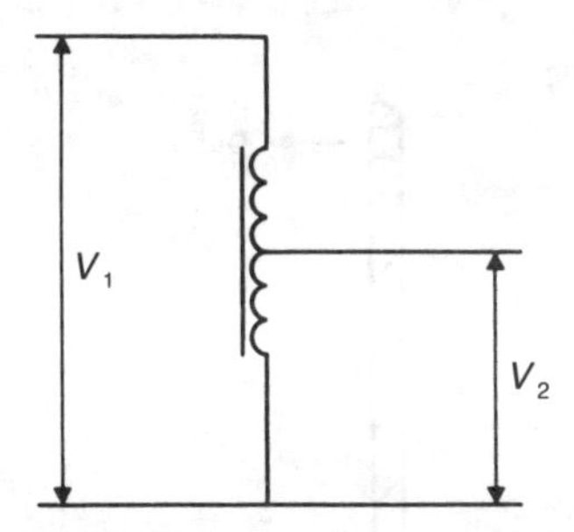

Figure 24.6. Diagram of auto-transformer.

minals of the transformer due to an internal fault, and, if the transformer is a step-down type, dangerous conditions could then exist. The transformer can be used to increase (step-up) or decrease (step-down) the voltage, but normally has a ratio of less than 2 to 1.

The construction is generally the same as the double-wound transformer with the same type of cooling. Regulations covering the use of this type of transformer are summarised at the end of this chapter.

Three-phase transformers

The majority of transformers used in industry and distribution are designed for three-phase operation. In the double-wound type, three separate transformers are wound on a common core. The primary windings are connected in either star or delta, and the secondaries connected likewise. In distribution transformers the secondary is commonly connected in star. This has the advantage of providing a fourth, or neutral, connection and enables the transformer to give *two* values of secondary voltage. Connecting between any of the three winding ends gives full line voltage, and connecting between the neutral and any winding end gives 58 per cent of the line voltage. A typical distribution transformer gives 415 V between the lines; connecting between line and neutral gives 58 per cent of this which is 240 V. The transformer can therefore be used to supply three-phase loads, such as motors, at 415 V, and lighting and heating loads at 240 V.

The construction is the same as that previously described, but because of the larger sizes used in the three-phase work, cooling is usually by oil. Three-phase auto-transformers are also available, and are mainly used for motor starting.

Applications

Transformers are used in every branch of the electrical industry ranging from domestic bell transformers to the large three-phase units used in generating stations. Single-phase, air-cooled types are used for bell circuits, radio receivers and discharge lighting. Isolated farmhouses and small housing schemes are usually fed from pole-mounted, single-phase, oil-cooled units. These latter transformers step-down from the rural distribution voltage of, say, 11 kV, to the required 240 V. Special high-voltage fuses are used to protect the transformer against faults.

The three-phase transformer is used widely in generation and distribution. They step up the generated voltage from 25 kV to 275 kV required for transmission and are used to reduce this high voltage to that for normal local distribution. The majority of factors are supplied at 11 kV or 33 kV and have their own transformers to give the required voltages for motors, heating and lighting.

Auto-transformers are used in some types of discharge lighting, but the three-phase types are usually confined to motor starting.

Installation and maintenance

The general precautions to be observed when installing transformers are the same as those for all electrical machinery. The apparatus must be installed to allow its safe operation and protected as far as possible from accidental damage.

Particular care must be taken in the installation of oil-filled transformers to minimise the danger of fire and explosion. A transformer containing more than 25 litres should be housed in a room of fireproof construction, e.g. brick. Where the quantity of oil exceeds 250 litres, a rubble-filled pit should be provided into which leaking or burning oil can drain. The room should have ventilators to the outside. In the case where several large units occupy the same room, blast walls are built between the transformers. Normally, the large transformers used in industry and distribution

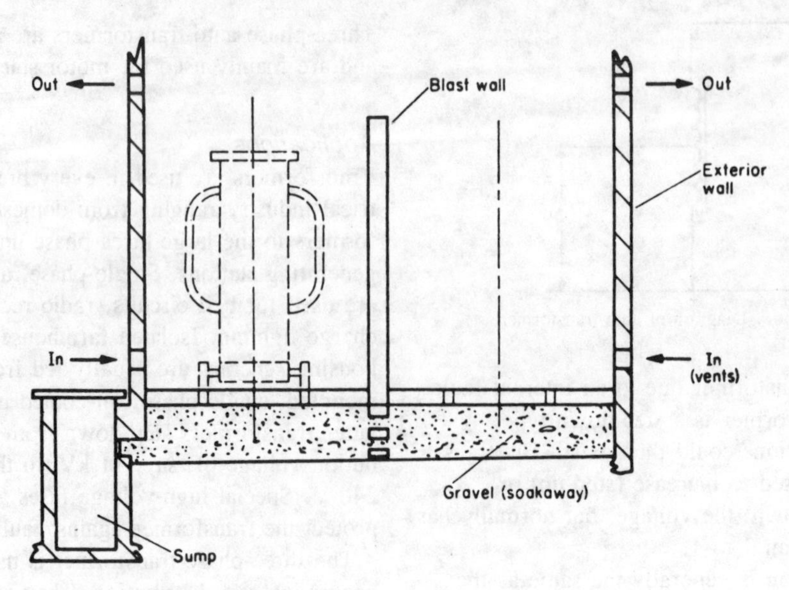

Figure 24.7. Site installation of an oil-filled transformer.

work are located outdoors in an annexe to the main substation.

All transformers should be checked at regular intervals for soundness of insulation, using a Megger to check both primary and secondary windings. The oil in oil-cooled units tends to become slightly acid in use and lose some of its insulating qualities. Samples of the oil are drawn off, say, every six months, and analysed for acid content and insulation strength. General maintenance includes ensuring that the level of oil is maintained, visual inspection of cable terminations, etc., and in the case of air-cooled units, checks for obstruction of ventilation louvres.

Regulations

The following is a brief summary of the regulations relating to electrical equipment. Reference should be made to the current edition of the IEE Regulations and to the various British Standards Codes of Practice which indicate accepted methods of maintenance.

1. Chokes and transformers, classified as fixed apparatus, exceeding 60 W capacity are to be adequately ventilated and enclosed in a proper container or so mounted to minimise fire risks. In the latter case, the mounting must be such that no wood, except hardwood, or combustible material is within 300 mm measured vertically above the apparatus or 150 mm in any other direction from it. Combustible material protected by asbestos or other fireproof material is accepted. The housing of a choke in a discharge-lamp fitting also must comply with the regulations.
2. A step-up transformer, forming part of a consumer's installation, must be provided with a multi-pole, linked switch to completely isolate the transformer from the supply.
3. Transformers containing more than 25 litres of oil must have facilities for draining away any excess oil. The oil must be prevented from leaking into parts of the building. This regulation is met by mounting the transformer over gravel-filled pits.
4. Buildings housing oil-filled transformers should be of fireproof construction. If the quantity of oil is more than 200 litres, the building should be ventilated to the outside.
5. Auto-transformers shall not be supplied from the mains where the voltage exceeds 250 V.

Exceptions are made if the transformer is used for motor-starting, or with a capacitor for improving power factor. The auto-transformer must be installed beside the capacitor.

Portable appliances, socket-outlets and extra-low voltage apparatus shall not be supplied from an auto-transformer. It should be noted that portable appliances include toys, model railways, etc.

25 Motors and control gear

General

Motors are the most common and best known of electrical machines. They are classified in two groups depending on whether they are suitable for use on direct-current or alternating-current systems. There is a third motor known as the Universal which can be used on dc or ac and these are used to drive small appliances such as vacuum cleaners, food mixers, etc. Each of the two main groups is subdivided giving several types of motor, each type having its own particular application.

More ac motors are in use than dc motors, and this is due to two factors. The majority of our supply systems are alternating current and allied to this is the fact that the ac motor is simpler and cheaper than its dc counterpart. Where variable speed over a wide range is required such as cranes, locomotives, etc, use is made of the direct-current motor, since the basic ac motor is essentially a single speed machine. This means that where the supply is alternating it is necessary to install rectifying equipment which adds to the cost of the installation.

The Regulations require that motors above 0.37 kW require control gear which incorporates a device to prevent self-starting and protection against overcurrent. Each type of motor requires its own particular form of control gear and this will be discussed at relevant points throughout the chapter.

Motor enclosures

Motors require some protection, both to prevent the user coming into contact with live and rotating parts and to protect the motor from damage due to dampness, dust, etc. All motors consist of a stationary part called either the yoke (dc) or the stator (ac) and a rotating part referred to as the armature (dc) or the rotor (ac).

In the case of very large motors driving machinery in a dry dust-free atmosphere the rotating part is often supported on pedestal bearings. The motor is then fenced off to prevent anyone from coming into contact with the conductors or moving parts. It is obvious that this method is only permissible in ideal atmospheric conditions and where no work is carried out near the motor. In other situations a more robust construction is required and details of various enclosures are given below.

(a) Screen-protected. The motor is enclosed in a steel casing and the shaft runs on bearings housed in the end covers of the casing. The end covers are fitted with metallic screens and an internal fan drives cooling air through the motor. This type of motor is used in dry, dust-free conditions and is the most common enclosure in use.

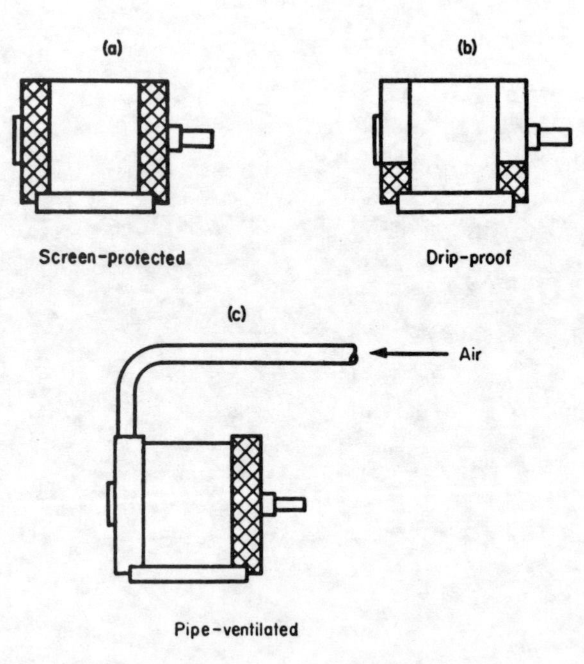

Figure 25.1. Types of motor enclosures.

(b) *Drip-proof.* This enclosure is similar to the screen-protected type, but the screens are fitted with angled cowls. These cowls prevent any dripping water from entering the machine. The motor can be used in damp conditions but it must be remembered that the enclosure is not waterproof. Again, the enclosure is not suitable for dusty atmospheres as particles of dust blocking the screens could cause overheating.

(c) *Pipe-ventilated.* The air which is circulated through the motor by the internal fan is brought from outside the building in which the motor is located. This type of enclosure is useful when the motor is located in very dusty situations such as woodworking shops.

(d) *Totally enclosed.* There are no air inlets in this type of enclosure, the cooling being by means of radiation from the surface of the motor. An internal fan is fitted and the casing is ribbed to increase the cooling surface. The motor runs hotter than any of those previously described and a high-class insulation material such as fibre glass is often used to withstand the higher temperatures. Dirt and moisture are excluded, and such an enclosure is suitable for use in corrosive atmospheres. Care must be taken that the ribs do not become clogged up or overheating will occur.

(e) *Flameproof.* In construction these are similar to the totally enclosed machines but are physically larger and more robust. The end covers are bolted to the main casing, which is fairly thick metal; and special seals are fitted at the shaft bearings. The construction has to be sufficiently strong to withstand an internal explosion without allowing the passage of a dangerous flame which could ignite the external atmosphere. These motors are used in the mining and oil industries or any situation where explosive atmospheres are encountered. Great care must be exercised in the maintenance of these machines.

In addition to the above general types, water-cooled motors are also available. Water is pumped through ducts in the stationary part and through the hollow shaft and along ducts in the rotating part. Several firms also make submersible motors which are completely watertight. These are used for driving pumps in wells and boreholes.

DC motors

This book does not go deeply into the theory of the electric motor: the following notes give only brief details of the action.

The motor basically consists of a set of coils wound on a laminated-steel core or armature. The armature is located on a shaft running on bearings and is free to rotate between the poles of a magnet. When current is passed through the coils a force is produced causing the armature to turn. The magnitude of the force depends on (*a*) field strength, (*b*) value of current and (*c*) length of coil.

In theory the coils will only turn until they lie at right-angles to the magnetic field, but by reversing the current when the coils are midway between the poles continuous rotation is produced. This reversal is achieved by using a commutator, which is a set of copper segments insulated from each other and from the motor shaft. The ends of each coil are connected to segments of the commutator and current led in and out of the coils by brushes running on the commutator. Figure 25.2 shows the

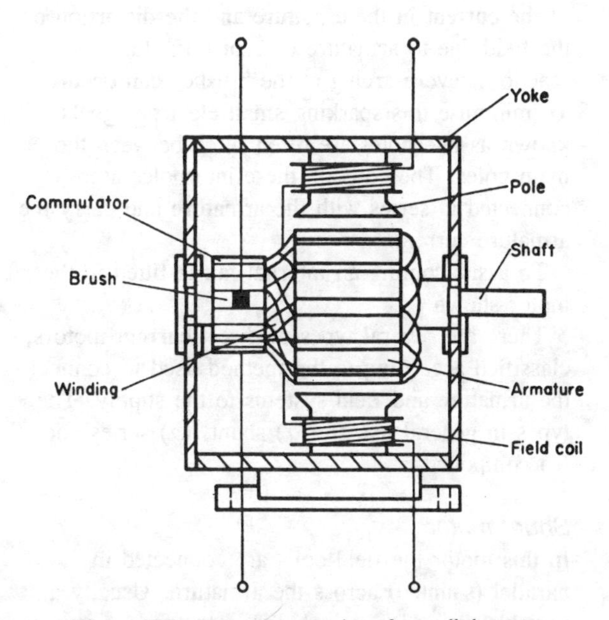

Figure 25.2. General construction of a small dc motor.

general construction of a small direct-current motor.

The coils are laid in slots on the armature and are held in place by wooden or fibre wedges. In the simple motor shown the coils are connected in series and the ends of each coil brought out to segments on the commutator. The carbon brushes are held in spring-loaded boxes and run on the commutator surface. In very large machines several brushes, connected in parallel, may be required to carry the heavy current involved. Ball and roller bearings, carried in the main casing or in external pedestals, support the shaft. Sleeve bearings, in which the shaft turns in cast-steel, white-metal lined, oil-lubricated sleeves, are often used in large machines. The shaft is machined and a keyway provided to allow the fitting of pulleys, couplings or gearing.

The stationary part, or yoke, carries the magnet poles and is also part of the magnetic circuit. In practice electromagnets are used and the strength of the magnetic field is controlled by varying the current flowing in the magnetising coils. It will be seen later that by varying the field current, which in a large machine is only a few amperes, the speed of the motor can be varied. The pole pieces and yoke are usually cast-steel and the poles are bolted to the yoke. Owing to the sudden reversal of the current in the armature and the distortion of the field due to armature current (armature reaction) severe arcing at the brushes can occur. To minimise this sparking small electromagnets known as interpoles are often fitted between the main poles. The coils on these interpoles are connected in series with the armature and carry the armature current.

To assist cooling an internal fan is fitted in the motor shown.

There are several types of direct-current motors, classified according to the method used to connect the armature and field systems to the supply. The types in general use are (*a*) shunt, (*b*) series and (*c*) compound.

Shunt motor

In this motor the field coils are connected in parallel (shunted) across the armature. Usually a variable resistance is included in the field circuit to control the field current and therefore the speed of the motor. The circuit diagram of a shunt motor is shown in Figure 25.3.

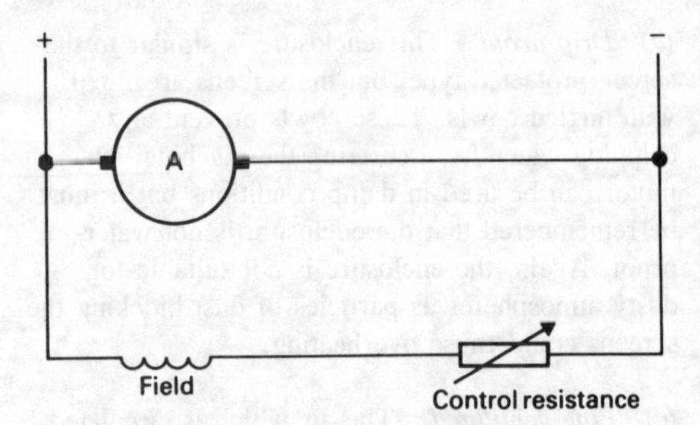

Figure 25.3. Circuit diagram of a dc shunt motor.

The speed of the shunt motor falls slightly as the load it is driving increases, but can be taken as fairly constant over the range between no load and full load: above full load the speed drops rapidly. The motor is thus suitable for driving machines such as pumps, lathes, etc., where sudden overloads do not occur. This is the most common dc motor.

Starting and speed control. All except the smallest of motors require some type of starter to prevent heavy current being drawn from the supply on starting. For dc machines the basic requirement is a graded resistance in series with the armature to limit the starting current. The resistance is cut out of the circuit in steps as the motor speeds up. To comply with the regulations, however, means to prevent self-starting after failure of supply, and overcurrent protection is added to the basic starter. The starter is required because on switching on the motor the current would be limited only by the armature resistance (about 1 ohm), only if no series resistance is used. As the conductors begin to turn an e.m.f. is generated in them in a direction opposite to the applied voltage. This e.m.f. is known as the 'back e.m.f.' and the voltage acting on the armature at any time is the difference between the applied voltage and the back e.m.f. The value of the back e.m.f. of a standard dc machine on normal duty is only a few volts less than the applied voltage. As an example the back e.m.f. of a 400 V motor may be 390 V, the

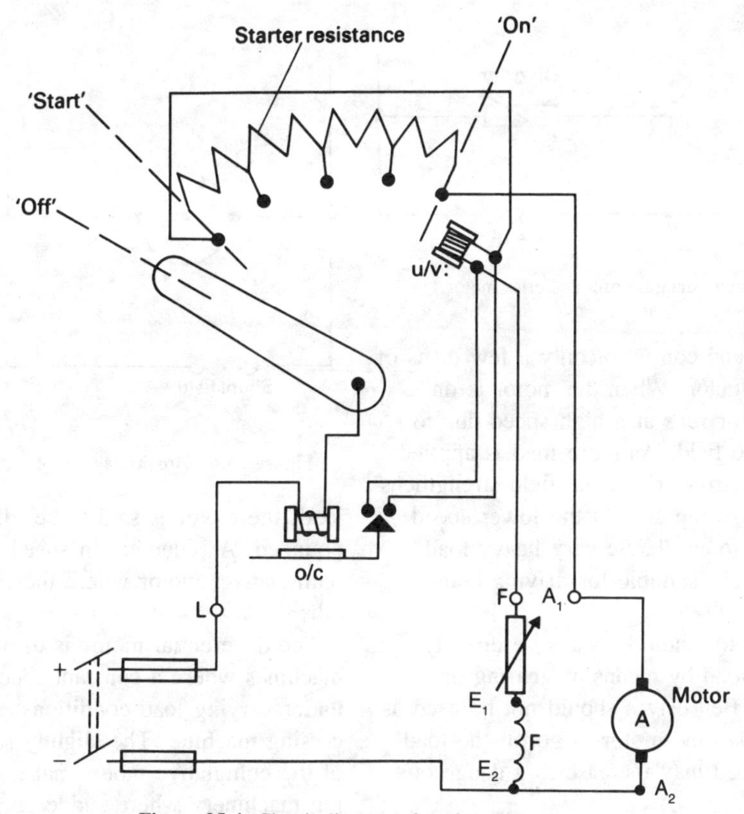

Figure 25.4. Circuit diagram of dc faceplate starter.

voltage acting on the armature being 400 V − 390 V = 10 V. The small resultant voltage is sufficient to circulate the armature current required to develop the correct horse-power. The starter is therefore cut out in steps as the back e.m.f. builds up with the increase in motor speed.

The faceplate starter is suitable for use with a shunt motor. The device to prevent self-starting is called the under-voltage coil and is an electromagnet through which the field current passes. The starter handle is attracted and held in place by this electromagnet when the starter is in the fully ON position. Should the supply fail, or the field circuit become broken, the coil is de-energised and the handle is pulled, by the spring, back to the OFF position.

Overcurrent protection is provided by a second electromagnet, the coil of which carries the motor current. If the current becomes excessive the trip-bar is attached to the electromagnet and the contacts close, short-circuiting the under-voltage coil. The starter handle is again released and returned to the OFF position.

It should be noted that the action of the starter is such that the field circuit is completed as the handle moves to the first stud of the starter. If the supply to the armature is maintained without the field circuit being complete, the motor, if running, will tend to speed up and may reach dangerously high speeds.

The speed of the motor can be changed by either varying the armature voltage or the field current. Invariably variation of the field current is used as only a few amperes flow in this circuit, whereas the armature current may be very high and any control gear would have to carry this large current, if armature voltage control were used. Reducing field current increases the speed.

Series motor

The armature and the field are connected in series as shown in Figure 25.5. The field coils carry the

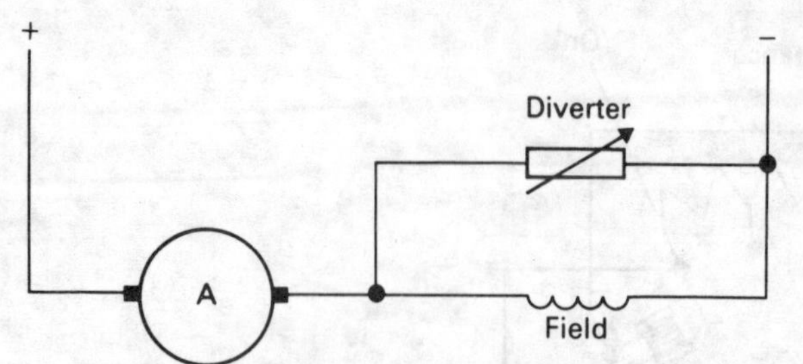

Figure 25.5. Circuit diagram of a dc series motor.

armature current and consist of only a few turns of heavy-gauge conductor. When the motor is on light load the motor runs at a high speed due to the weak magnetic field. As more load is applied to the motor the current rises, the field strengthens and the speed drops rapidly. At the lower speed the motor is able to accelerate very heavy loads and is therefore very suitable for driving cranes, electric locomotives, etc.

This type of motor should always be directly connected to the load by means of gearing or bolted couplings. Belt drives should not be used as a broken belt causes the motor to go on 'no-load' and the motor speed may increase to a dangerous level.

Starting and speed control. The starter in this case is a high-rated resistance in series with the armature. Overcurrent and under-voltage protection is provided as in the shunt-motor starter. Speed control is effected in traction and hoist work by using the heavily rated starting resistance as a controller to vary the voltage across the motor terminals. A degree of speed control is achieved by by-passing some of the field current through a 'diverter' shunted across the field.

Compound motor

This motor has both shunt and series fields as shown in Figure 25.6. The fields are connected so that they may assist each other, strengthening the field with increased load, or oppose each other. Both field coil systems are wound on the same motor poles; the degree of 'compounding' causes the motor speed to vary with load in a particular way. The speed can be made to rise, remain constant or fall with increase in load. If the speed rises slightly, or stays constant with increase in

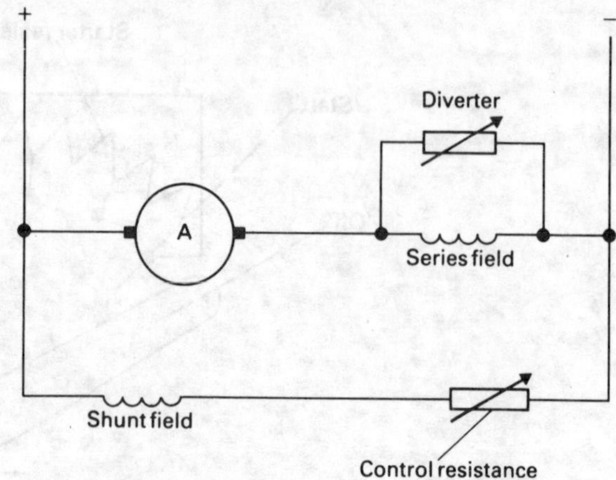

Figure 25.6. Circuit diagram of a dc compound motor.

load, the motor is said to be 'differentially' compounded. A reduction in speed is obtained from a 'cumulative' motor where the fields assist each other.

The differential motor is useful for driving machines where a constant speed is necessary under varying load conditions, such as a processing machine. The slightly series characteristic of the cumulative motor makes it suitable for driving machinery where sudden heavy loads are imposed for short periods. An example of this is a strip rolling mill where ingots of steel are passed through rollers to be rolled into steel plates. This motor does not suffer from the disadvantage of series motors that it will overspeed if the load is removed.

Starting and speed control. The compound motor is started using the same faceplate starter as already described for the shunt motor. Speed control is effected by a variable resistor in the low-current shunt field circuit.

Reversal of dc motors

A dc motor is reversed by changing over the connections to *either* the armature or field circuits. Reversal of the mains supply leads will cause the motor to rotate in the same direction as before. Where a reversing switch is used this is always connected in the shunt field circuit in shunt and compound motors. The reason for this is that these fields carry very small currents relative to the armature.

AC motors

Most of the motors used in industry and the home are of the alternating-current pattern. Industrial motors are usually designed for three-phase operation and the domestic motors such as used on washing machines, etc. are designed for the single phase systems used in the home. The size of single phase motors used in industry seldom exceed 7 kW.

Three-phase motors are fairly simple in construction and have the advantage of being self-starting. The single-phase motor is not self-starting and requires to be wound in a special way and fitted with special starting arrangements. This makes it much more expensive and larger than a three-phase motor of equivalent rating.

One minor disadvantage of the ac motor is that it runs at a fixed speed in standard motors. The speed depends on the number of poles and the frequency of the supply system. The relationship between the speed, poles and frequency is given by the formula $N = 60f/p$ where N is the approximate speed in revs/min., f the frequency in Hertz and p the pairs of poles in the motor. As an example the speed of a four-pole motor operating on a 50 Hz supply system is $60 \times 50/2 =$ 1500 rev/min approximately.

The speed of some industrial motors is changed by 'pole-changing' in which the motor windings are switched to give different numbers of poles. It should be noted that this system only gives different values of fixed speeds. Variable-speed commutator ac motors are also available but these are out of the scope of this chapter.

The motor consists of a fixed portion, the stator, which is built up of laminated steel plates. Laminations are used, rather than a solid core, to minimise the eddy current losses associated with ac magnetic circuits. The stator carries the field winding which is distributed in slots around the core, and the winding is retained in the slots by wooden or fibre wedges.

The rotating part, rotor, is made up of laminated plates, mounted on the motor shaft, and the rotor can be of the squirrel-cage or wound-rotor design. Both types of rotor are shown in Figure 25.7. A squirrel-cage rotor has lightly insulated copper or alloy bars laid in slots around the rotor. These bars are short-circuited at each end of the rotor. This is the most common type of rotor construction in use. In the wound rotor, coils are laid in the rotor slots and the coil ends connected to slip-rings on the motor shaft. Carbon, or composition, brushes running on the slip-rings allow the rotor to be connected to an external, variable, resistance for starting purposes.

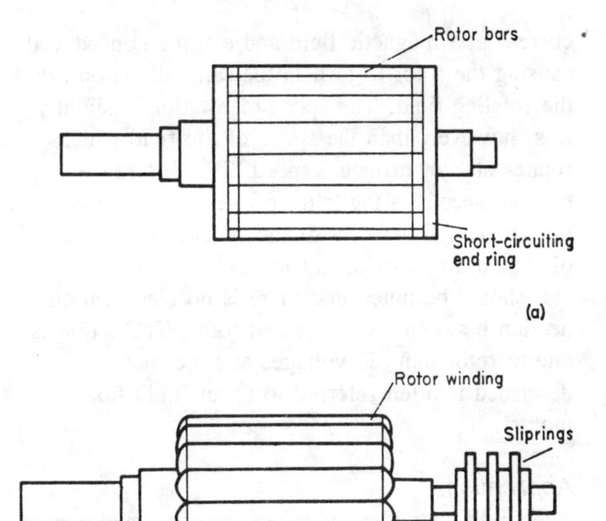

Figure 25.7. (a) Squirrel-cage rotor of ac induction motor; (b) Wound rotor of ac induction motor.

Three-phase motors

The basic principle of these motors is as follows: three sets of stator coils are uniformly distributed around the stator core and connected in either star or delta. The rotor can be of the squirrel-cage or wound-rotor type and if the latter is used the three rotor windings are usually in star and brought out to three slip-rings.

When the stator is connected to the ac supply, the field produced rotates around the stator at a speed given by the formula previously quoted ($N = 60f/p$). The machine behaves like a transformer where the stator is the primary and the rotor the secondary. The ac applied to the stator causes an alternating voltage to be induced in the rotor and a current therefore circulates in the short-circuited rotor winding. The rotor conductors now carry current in a magnetic field and a force is produced causing the rotor to turn in the same direction as the rotating field. The speed of rotation is slightly less, however, than the speed of the field which rotates at 'synchronous' speed. The difference in the two speeds is the 'slip' of the motor and is of the order of 5 per cent of the synchronous speed of standard motors at full load.

It should be noted that there is no electrical connection between the stator and rotor. The action is due to rotor induced voltages and the motor described is often referred to as an 'induction motor'.

Applications

The speed characteristic of an induction motor is similar to that of the shunt motor, the speed being fairly constant up to full load. Squirrel-cage motors are used for driving machines having a steady load such as workshop lathes, centrifugal pumps, air compressors and other machines which can be started on light load. Where the supply system can withstand the starting current surges, squirrel-cage motors of up to 2,000 kW and 11,000 V can be used. Normally, however, the rating of squirrel-cage motors rarely exceeds 75 kW where the industrial consumer is supplied from a medium sized substation, at 3,300 V and 415 V.

Slip-ring motors are used where heavy machinery has to accelerate or where the machine is started against full load. Examples of these conditions are hoists, mining haulages, large ventilating fans, etc.

Control gear

Small squirrel-cage motors, starting on light load, can be switched direct-on-line, but to comply with the IEE Regulations the starters are fitted with isolator, and overcurrent and under-voltage protection. On starting, the motor, if switched direct, takes a current of up to seven times its normal full load current, which decreases as full speed is reached. The supply authorities require that the magnitude and duration of these starting surges be limited and use is often made of special starting gear.

Starting equipment for wound-rotor (slip-ring) motors comprises two parts: a stator switch, which contains the protective devices, and a graded rotor resistance which is cut out as the motor speeds up.

(a) Direct-on-line starting. This can be a hand-operated switch, but more use is now being made of the push-button operated contactor starter. Figure 25.8 shows the circuit diagram of a simple contactor starter.

The start button, when pressed, completes the contactor circuit and closes the three-phase switch establishing the supply to the stator. The motor current passes through the overload coils and if the current becomes excessive the overcurrent solenoid operates the trip coil and breaks the contactor circuit. Undervoltage protection is inherent, as failure of the supply causes the contactor, which is operated from the supply, to drop out. The stop-button is wired in series with the overload contacts to break the circuit. An isolating switch is mounted in, or near to, the starter so that the circuit can be isolated, and if necessary this can be locked off for safety during motor maintenance.

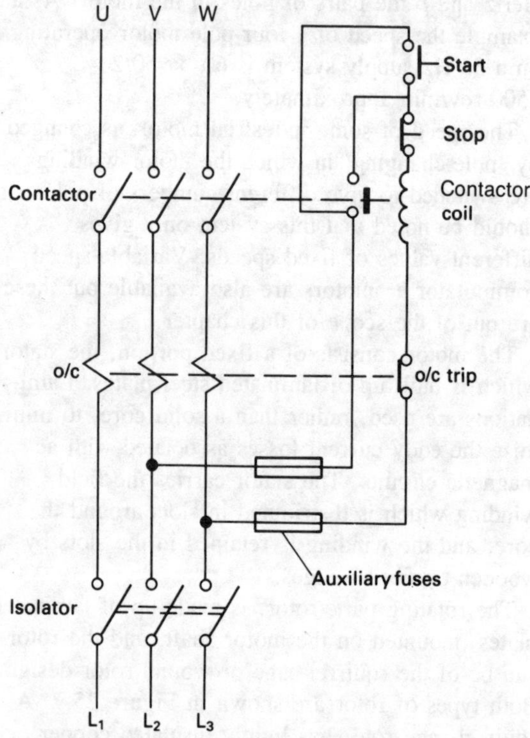

Figure 25.8. Circuit diagram of a contactor starter.

If a hand-operated starter is used the overcurrent device actuates the mechanical linkage of the starter. In this case, the undervoltage coil is energised from two of the supply terminals and also actuates the mechanical linkage if the supply fails.

(b) Star-delta starting. This method of starting requires that the two ends of each of the three stator coils are brought out to a six-terminal box. The motor windings are connected in star at starting which means that each winding receives approximately 58 per cent of the supply voltage. When the motor approaches its full speed the windings are switched over to give a delta connection in the full supply voltage applied to each winding and the motor develops its full horsepower. The current taken when the windings are in star is reduced, but the horsepower the motor can develop is only one-third that obtainable in delta. The method is only suitable, therefore, where the load being driven is reasonably easy to accelerate. Figure 25.9 shows the circuit diagram of a typical hand-operated star-delta starter. The drawing shows the devices required to comply with the regulations. Greater use is being made of automatic push-button starters using contactors. A timing device is used so that one contactor closes at starting for the star connection then after a pre-set time a second contactor closes to connect the windings in delta.

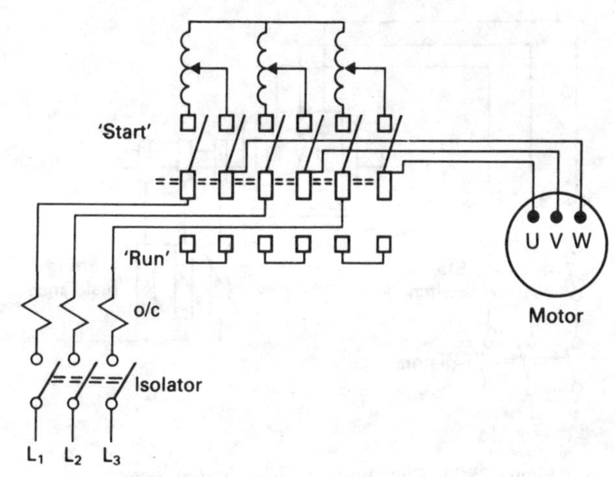

Figure 25.9. Circuit diagram of a star/delta starter.

(c) Auto-transformer starting. Where the reduction in horsepower output caused by the star-delta method of starting is too great use is made of the auto-transformer starter. In this case the motor is supplied from an auto-transformer which is tapped to give a reduced voltage on starting and then switched to give full voltage as the motor runs up to speed. Typical values of starting voltage are 70 to 80 per cent supply voltage. Figure 25.10 gives circuit details of the starter.

(d) Rotor starter. This method is used for slip-ring motors and the circuit is given in Figure 25.11. To avoid very high surges on starting the resistance must be 'all-in' before closing the stator

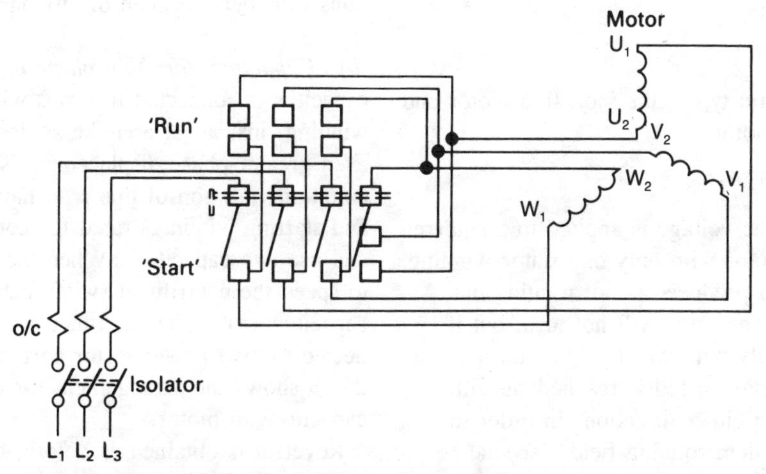

Figure 25.10. Circuit diagram of auto-transformer starter.

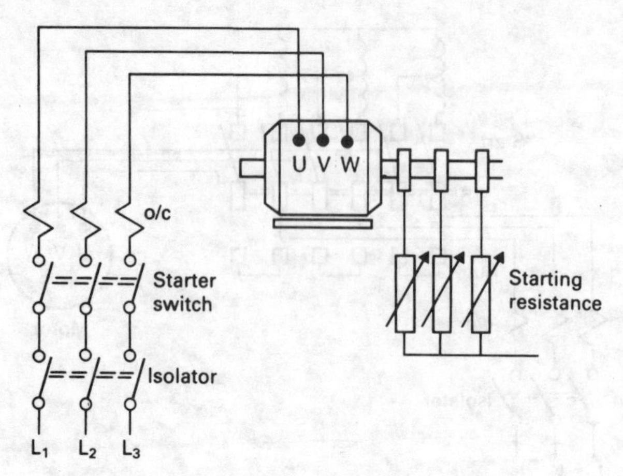

Figure 25.11. Circuit diagram of resistance starter.

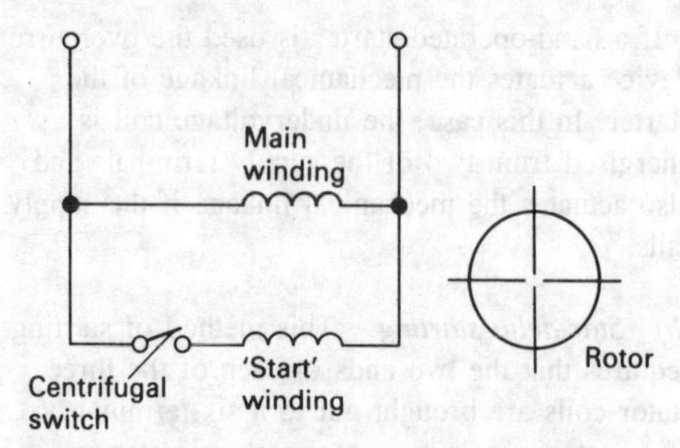

Figure 25.12. Circuit diagram of split-phase ac motor.

switch. A mechanical interlock is usually fitted so that when the stator switch is switched off, the rotor-resistance arm is automatically returned to the 'all-in' position, ready for the next time the motor is to be started. The rotor resistance gives some degree of speed control, on load, by switching out portions of the resistance only to give a speed lower than full speed. The resistance used for speed control must be very heavily rated to dissipate the heat generated by the rotor currents, and even then can only be used for short periods of time.

Again, an automatic starter is available, using timers and contactors to cut out sections of the resistance at pre-determined intervals.

Reversal of three-phase motors is obtained by changing over any two of the supply leads.

Single-phase motors

There are two main types, the induction motor and the commutator motor.

Induction motor

If a single-phase ac voltage is applied to a squirrel-cage induction motor with only one stator winding, the magnetic field produced is not rotating but merely pulsates. The rotor will not turn, but if turned mechanically will start to rotate on its own accord after a certain speed is reached; it will, however, rotate in either direction. In order to produce an equivalent rotating field a second winding, the starting winding, is also wound on the stator at 90° to the main winding. This winding produces a field which is out of phase with that due to the main winding and several methods are employed to obtain the phase displacement. Motors constructed in this fashion are called 'split-phase' induction motors. To minimise losses, and keep the power factor as high as possible, a centrifugal switch is fitted on the motor shaft which open-circuits the starting winding when the motor attains about 80 per cent of full speed.

(a) Inductor-start split-phase motor. The starting winding in this motor is of higher inductance than the main winding so causing phase displacement at starting. To obtain even greater displacement, and better starting characteristics with reduced current, a choke may be placed in series with the starting winding. Figure 25.12 shows the connection arrangements at start and when running. The motor is reversed by changing over the connections to either the main or auxiliary windings.

(b) Capacitor-start split-phase motor. If a capacitor is connected in series with the starting winding, instead of a choke as described above, an even greater phase displacement is obtained. A common variation of this is to have both the main and starting windings rated for continuous running and use two capacitors. When the motor runs up to speed the centrifugal switch cuts out one capacitor and leaves the other permanently connected to give power factor correction. Figure 25.13 shows the arrangement for a capacitor-start, capacitor-run motor.

Reversal is obtained as before, by changing connections to either main or auxiliary windings.

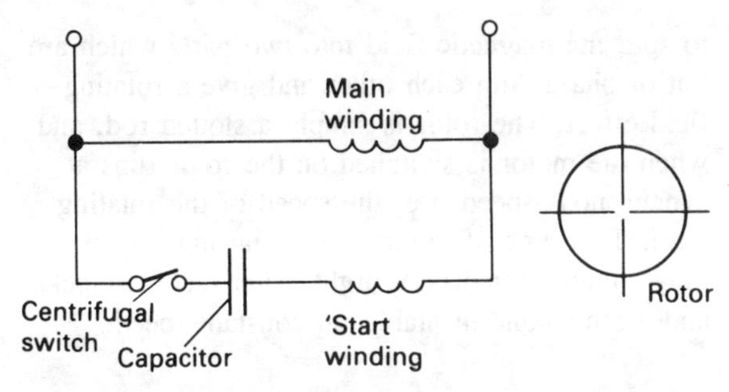

Figure 25.13. Circuit diagram of capacitor-start ac motor.

Both motors have speed characteristics similar to dc shunt motors and are used for driving workshop machinery and the larger domestic equipment such as refrigerator drives.

Starting is usually direct-on-line for the smaller machines or a series resistance can be used in the stator circuit and cut out as the motor speeds up. The majority of split-phase motors have squirrel-cage rotors, but wound rotor types are available for heavier starting duties. The wound rotor can be exactly as that used in a three-phase machine, and the three slip-rings are connected to a starting resistance.

Commutator motor

The motors described here are single-speed machines, although it must be remembered that special variable-speed commutator motors for use on ac are available.

(a) Single-phase series motor. As stated in the section on direct-current motors, reversal of the dc mains causes the motor to run in the same direction. If a dc motor were used on an alternating supply the current in the armature and field would reverse simultaneously. This would mean that the motor would run in one direction, and series motors are used on ac. These are often referred to as Universal motors since they are suitable for ac/dc working. When used on ac there is some trouble with commutation which, in a large motor, could give rise to severe arcing at the brushes. The motors are used only in small sizes and find their application in various domestic appliances. The speed characteristic is that of the dc series machine, the speed falling with increased

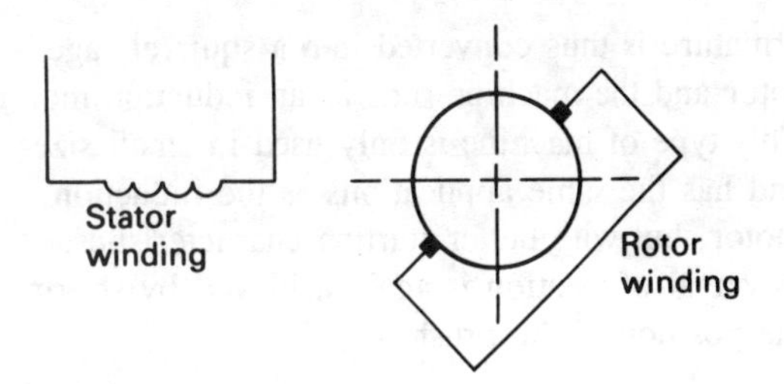

Figure 25.14. Circuit diagram of single-phase repulsion motor.

mechanical load. They are used to drive articles having a fairly constant load such as hand tools, vacuum cleaners, etc. To minimise losses when used on ac the stator (yoke) and rotor (armature) are built up of laminated steel, but otherwise the machine is the same as a dc motor.

Direct-on starting is usual for these small motors. Reversal of rotation is obtained as in the dc case, that is, reverse either field or armature winding.

(b) Single-phase repulsion motor. This consists of a single-winding on the stator and a commutator armature. Short-circuited sets of brushes run on the commutator surface. When the motor is switched on, the induced e.m.f. in the armature coils causes a current to flow in the armature. In order that rotation should take place, the brushes have to be moved round the commutator until their axes are at an angle to the stator field as in Figure 25.14. Interaction between the fields due to stator and armature currents then produce rotation. The speed of the repulsion motor behaves as that of a series motor, but the motor is capable of starting fairly heavy loads, and is available in much larger sizes than the series motor. The most suitable application is for constant speed heavy drives. Starting may be direct-on or by means of a series stator resistance. The direction of rotation is reversed by moving the brushes round the surface of the commutator.

(c) Single-phase, repulsion-start, induction motor. The construction is practically the same as the basic repulsion motor. The motor is started as before, but as the speed increases a centrifugal mechanism causes a ring to short-circuit the whole of the commutator and lift the brushes clear. The

armature is thus converted into a squirrel-cage rotor and the machine runs as an induction motor. This type of machine is only used in small sizes and has the same applications as the induction motor, but with better starting characteristics. Reversal of rotation is again achieved by shifting the position of the brushes.

(d) Single-phase, repulsion-induction motor. This machine has similar characteristics and applications as the repulsion-start, induction motor. The armature has a double winding, an outer commutator winding, and a squirrel-cage inner winding in the same slots. The motor starts as a repulsion type, but runs as an induction motor. There is no need for a short-circuiting and brush-lifting device; the motor has a fairly good power factor. Starting and reversal of rotation is as before.

Single-phase synchronous motor

These motors are used to drive clocks and other accurate timing devices. The speed is dependent on supply frequency which is maintained normally at a very accurate figure by the supply authorities. One form of synchronous motor is that known as the 'shaded-pole' motor. This motor consists of a horseshoe-shaped, laminated iron core on which the magnetising coil is wound. A copper ring is embedded on the face of each of the pole-pieces as shown in Figure 25.15. The effect of this ring is to split the magnetic field into two parts which are out of phase with each other and give a rotating field effect. The rotor is simply a slotted rod, and when the motor is switched on the rotor runs at synchronous speed, i.e. the speed of the rotating field. This type of motor is self-starting, but is only suitable for driving light loads such as clocks and gramophone turntables at constant speed.

Installation of motors

Serious consideration must be given to the type of motor enclosure used in a particular situation. Before the motor is installed an insulation test should be made. If this is low the motor should be stored in a warm place to drive out any moisture. Another method of drying out is to lock the armature or rotor and pass a current from a low-voltage supply through the windings; but this must be used carefully to avoid overheating.

Motors are usually mounted on steelwork firmly secured in a concrete base. Where pulley drives are used some method of belt tensioning is required and the motor is fixed on slide rails as shown in Figure 25.16. Adjusting screws bearing on the motor feet between motor and load are used for tensioning. The pulleys on the motor and the driven load must be exactly in line and parallel to each other. This is best checked by lining up the pulleys by a straight edge or line and Figure 25.17 shows the method of alignment.

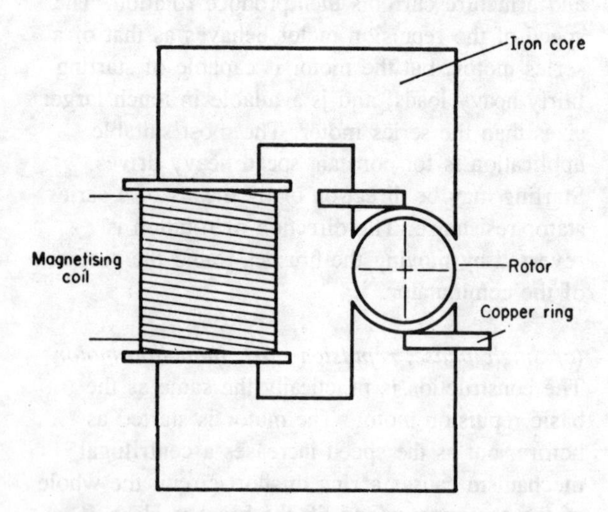

Figure 25.15. Diagram of single-phase synchronous motor.

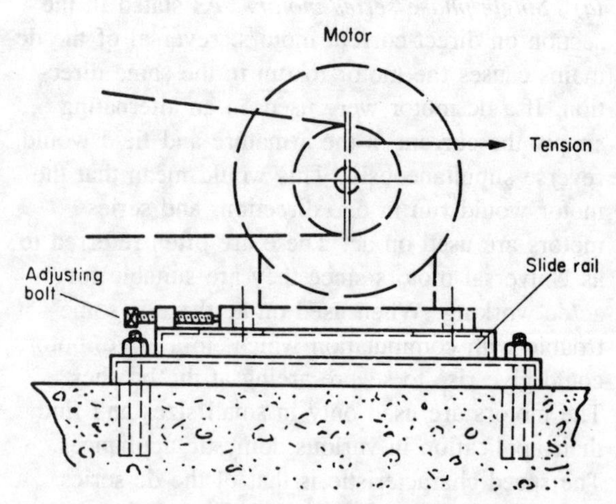

Figure 25.16. Motor mounted on slide rails.

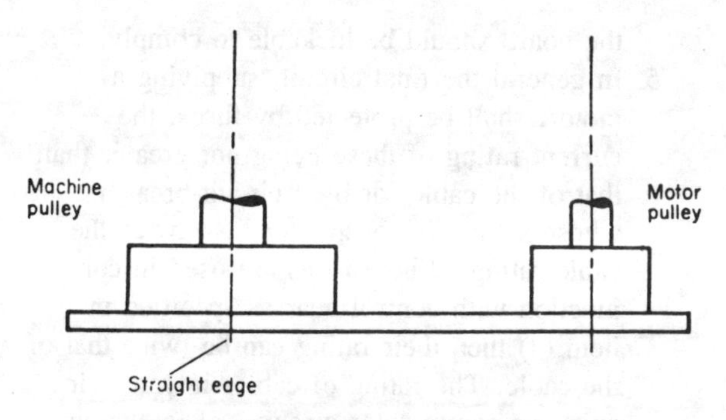

Figure 25.17. Lining up motor pulleys.

Pulleys and gearing are retained on the shaft by means of a key. The key should be of the drilled and tapped type which can be easily withdrawn using extracting gear. There should be no movement of the pulley on the shaft with a correctly fitted key, but it is necessary to have a little top clearance between the key and the pulley hub. Keys should never project beyond the end of the motor shaft. Where the motor is fitted on slide rails for adjustment, the cabling should be terminated in such a way that it allows movement of the motor. When armoured cables are used this presents little problem, but conduit installations require flexible extensions at the motor. Mineral-insulated cables are installed with a loop of cable at the motor termination point.

When the motor is ready to be put into service a final insulation test of the complete installation should be made. The bearings should be checked to ensure they are charged with a sufficient quantity of grease or oil.

Summary of applications

Motor type	*Application*
dc shunt	General purpose; constant load, machine drives.
dc series	Starting against heavy load, e.g. traction.
dc compound	
(*a*) cumulative	Fluctuating heavy loads, e.g. rolling mills.
(*b*) differential	Constant speed, e.g. processing.
3-Phase, squirrel-cage	General purpose. Light-load starting, e.g. machine drives.
3-Phase, slip-ring	Accelerating heavy loads, e.g. hoists.
1-Phase, induction	As 3-phase, squirrel-cage, but limited to about 3 kW.
Repulsion	As induction, but capable of starting heavier loads.
Universal	Small domestic appliances; hand tools, usually about 75 W.
1-Phase, synchronous	Timing equipment; Record-players and tape-recorders.

Summary of Regulations

The Regulations cover both dc and ac machines and lay down requirements regarding housing and construction of machines, provision of control gear, rating of cables and permissible volt drop in motor circuits.

The Regulations are summarised as follows:

1. The volt drop between consumer's terminals and motor terminals must not exceed 4 per cent of the declared or nominal voltage.
2. Where there is a danger of fire or explosion, the motor and its control gear shall be of flameproof construction. Alternatively, in dust-laden situations, the motor enclosure shall be of a type to exclude dust.
3. Motors used in situations where the temperature of the surrounding atmosphere is such that overheating could occur, shall be down-rated or the insulation shall be of a special material. Means for forced ventilation, or pipe-ventilated enclosures, can be used as an alternative.
4. Every electrical motor shall be provided with efficient means of starting and stopping. This control shall be within easy reach of the operator. In addition, motors of more than 0.37 kW shall be provided with control gear with the following devices: (*a*) means to

prevent self-starting in event of restoration of supply after failure, (*b*) protection against overcurrent in the motor and in the cables supplying the motor and (*c*) means of isolating the motor and all its associated apparatus.

In modern factories use is made of group control whereby several motors are automatically started from a group of contactors on a single board. It is useful to have a stop-button close to each motor position. Normally this stop-button is of the lockable type and, in addition, an isolator can be provided at the motor position. The main isolator on the board should be lockable to comply.

5. In general the final circuit, supplying a motor, shall be protected by fuses, the current rating of these being not greater than that of the cable, or by a circuit-breaker whose setting is not greater than twice the cable rating. Where fuses are used in conjunction with control gear as specified in item (4) then their rating can be twice that of the cable. The rating of cables in rotor circuits, or commutator circuits, of induction motors must be suitable for starting and running conditions.

26 Capacitors and power factor

Capacitors are used in installation work mainly for power factor correction, starting of single-phase motors and anti-interference devices. A brief description of the action of capacitors and the effect and importance of power factor is given here. For fuller treatment of both subjects, a good electrical science book should be consulted.

Capacitor action

The basic capacitor consists of two plates, separated from each other by a layer of air or an insulating material known as the 'dielectric'. If this arrangement is connected to a dc source, electrons leave one plate and travel round the circuit to the other plate. During this period a current flows for a short time before falling to zero. An excess of electrons on one plate and a deficiency on the other causes a potential difference (voltage) to exist between the plates. When this voltage is equal and opposite to the applied voltage, the current flow ceases. The capacitor is then 'charged' and will remain so even when the external supply voltage is removed. Connecting the terminals of the capacitor to another circuit will cause a current to flow until the potential difference between the plates falls to zero. The capacitor is then 'discharged' and the electrons rearrange themselves to give this condition. Figure 26.1 shows the action of a capacitor in a dc circuit.

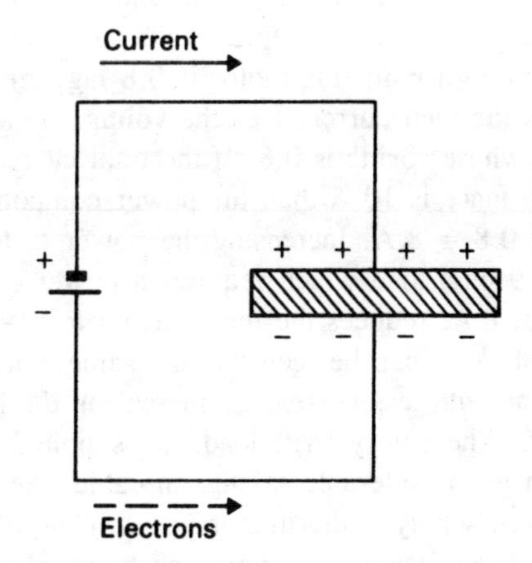

Figure 26.1. Action of capacitor in dc circuit.

If the capacitor is now connected to an alternating voltage source the conditions are changed. The capacitor charges as the voltage rises from zero during the first quarter cycle, and discharges as the voltage returns to zero during the next quarter cycle. The process is repeated throughout the next half cycle, but in the reverse direction. It can be seen that steady conditions never exist and an alternating current, of the same frequency as the supply, flows. This current, however, leads the applied voltage by 90°, that is by a quarter cycle. During the operation ac electrostatic fields are produced which cause energy to surge to and from the capacitor in alternate quarter cycles.

Power factor

Most of the machines used in electrical work are inductive (e.g. motors, transformers). An inductive circuit is one which possesses resistance and inductance. In such circuits the current lags behind the applied voltage by an angle ($\theta°$) which is between zero and 90°. The current has two components, the power component, in phase with the voltage, which does useful work and the 'idle' or reactive component 90° out of phase with the voltage. This idle component is responsible for the setting up of magnetic fields, causing energy to surge to and from the load during alternate quarter cycles. It should be noted that this is similar to the energy in a capacitor. If an inductor and capacitor are connected to the same ac source, the inductor would take in energy, as the capacitor was giving out energy (discharging).

The power component of the current is responsible for doing useful work (e.g. developing the machine output) and is constant for a given load. If the angle between the total current and voltage is reduced, without altering the power component, the total current for a given duty is reduced. The ratio of the power component to total current is the 'power factor' of the circuit and is a figure normally less than 1. Typical values of power factor for an induction motor is 0.8 lag; this means the total current lags the voltage by an angle whose cosine is 0.8. If the total current in this instance is 10 A then the power component is $10 \times 0.8 = 8$ A. Increasing the power factor to say 0.9 on a load which requires a power component of 8 A, reduces the total current to $8 \div 0.9 = 8.88$ A. It can be seen that the same work can be done with less current by increasing the power factor. Where very large loads are supplied this can mean considerable savings in cable size. In addition, supply authorities impose a charge for poor power-factor; and energy costs are also reduced by improving the power-factor. Figure 26.2 gives the phasor diagram before and after correction.

The most common method of improving the power factor of a machine is by connecting a capacitor in parallel with the machine. The capacitor takes a leading current; this compensates for and reduces the idle component of the machine current. The power factor is therefore increased and the total current reduced. Care must be taken that the capacitor used is not too large or the total current will lead the voltage. This would produce an idle component as before and increase the

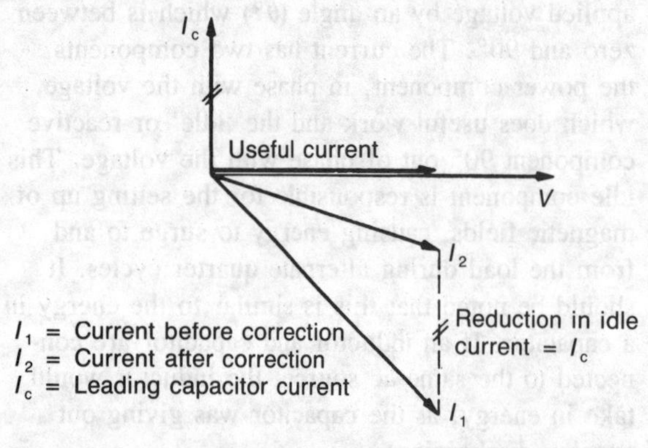

Figure 26.2. Phasor diagram of circuit before and after power-factor correction.

current. Over-correction of the power factor is sometimes avoided by use of a device to automatically cut out the capacitor when the machine is on light load.

Capacitor construction

Most of the capacitors used in installation work are metal-foil, paper-insulated. The capacitance depends on the area of the plate and on the thickness and the insulating qualities of the dielectric. A common method of increasing the area of the plates is to use two metal-foil strips separated by wax- or oil-impregnated paper. The physical size is kept down by rolling the plates and paper to give a tubular capacitor. The whole assembly is then housed in a plastic container, although higher voltage units and larger sizes are usually housed in a steel casing. In smaller sizes this is the type used for single-phase motor starting and power-factor correction in discharge-lamp circuits.

Where the power factor of large motors is to be improved, the paper capacitor is again used. In order to give the required capacity several capacitors are made up in banks. The banks are placed in a steel casing which is then filled with insulating oil in the same way as a transformer. If the motor, or load, is three-phase, three banks are used and connected in delta or star as required. The three banks of capacitor normally share the same tank.

As mentioned earlier the capacitor remains charged when removed from the supply, and, depending on the quality of insulation, can remain charged for a considerable time. To avoid danger of shock when working on a capacitor discharge resistors are fitted to the larger sizes. These resistors are permanently connected across the capacitor terminals and are of a high value, usually several megohms. They allow any charge on a capacitor to be safely dissipated after the capacitor is switched off. Care must be taken during capacitor maintenance not to remove these resistors. Before working on a power factor correction capacitor, at least ten minutes should be allowed, after switching off, to allow the capacitor to discharge.

The installation and maintenance of oil-filled capacitors is generally the same as that of transformers. Fire and explosion precautions must be taken and periodic checks on oil samples made. The capacitors used on lamps and motors are connected across the terminals of the apparatus (i.e. in parallel). Correction of an entire system (e.g. of a factory) is sometimes done by installing large capacitors in the main substation. In this case, each capacitor is controlled by a circuit-breaker, with protection against overload of cables and apparatus faults.

Capacitors used in radio work are sometimes of the multi-plate or electrolytic type. The multi-plate construction consists of a set of fixed and a set of moving plates. The capacitance is altered by moving the moving plates in or out of mesh with the fixed plates. The insulation in this case is air. The electrolytic capacitor is similar in appearance to the tubular paper capacitor. The plates are formed by electrolytic action which takes place when the capacitor is connected to the supply. This type of capacitor must be connected in the circuit with the correct polarity or damage will result.

A further use of capacitors in installation work is to act as a suppression device on an inductive circuit. The capacitor is connected across the contacts of a switch, such as the make-and-break of a trembler bell. The energy of the spark is then used up in charging the capacitor. In the section on lighting, the diagrams show a capacitor across the starting switch. This acts as a suppressor and prevents radio interference.

Regulations

The general regulations covering the installation and operation of electrical machines and apparatus also apply to capacitors. Precautions must be taken against overheating and fire risk. Every capacitor, except radio interference types, must be provided with means of automatic discharge. Small capacitors are exempt from this requirement. Plugs containing capacitors for radio interference must comply with the requirements of BS 613. It is suggested that the capacitor should be fitted to the appliance rather than the plug.

27 Circuit and wiring diagrams

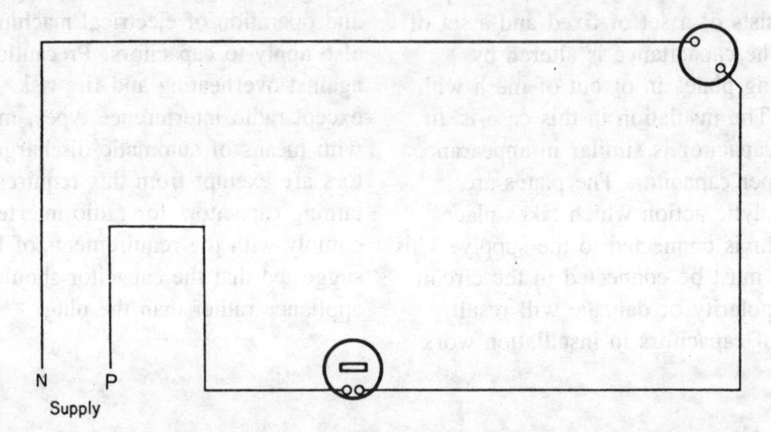

Figure 27.1. Wiring diagram of a lamp controlled by a one-way single-pole switch.

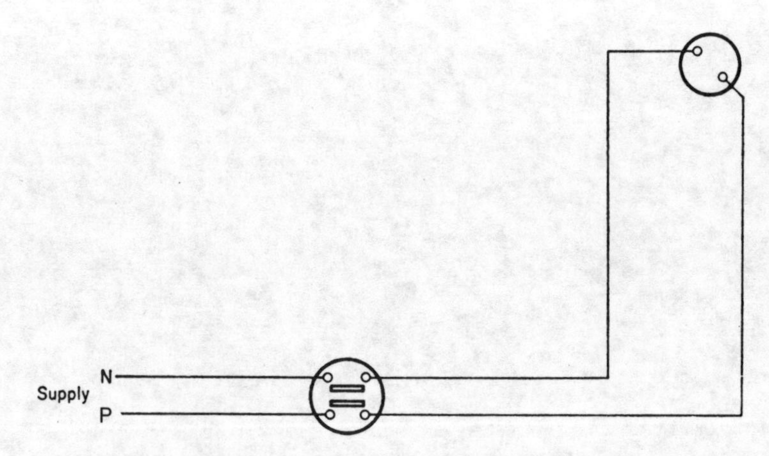

Figure 27.2. Wiring diagram of a lamp controlled by a double-pole switch.

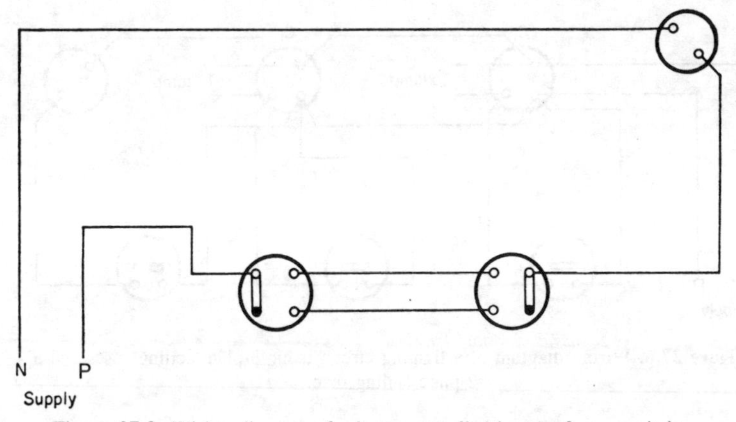

Figure 27.3. Wiring diagram of a lamp controlled by two 2-way switches.

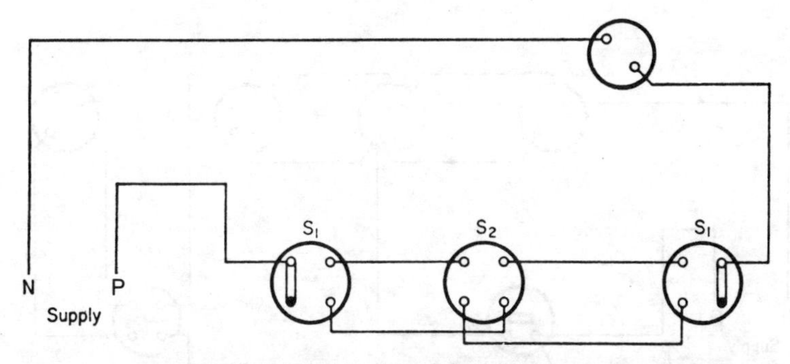

Figure 27.4. Wiring diagram of a lamp controlled by an intermediate switching arrangement with two 2-way switches and one intermediate switch.

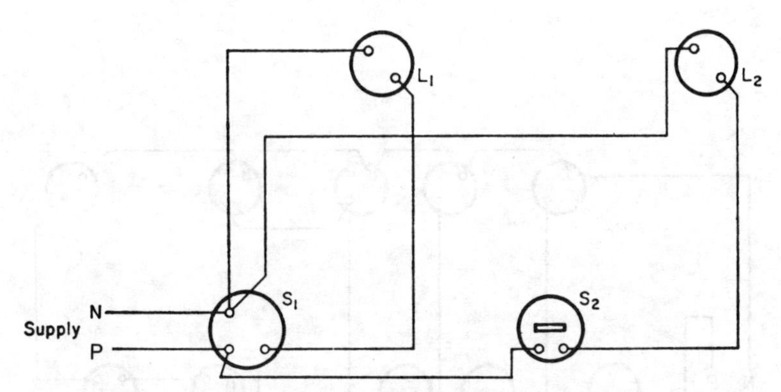

Figure 27.5. Wiring diagram of one lamp controlled by a one-way switch with a loop-in terminal and one lamp controlled by a one-way switch.

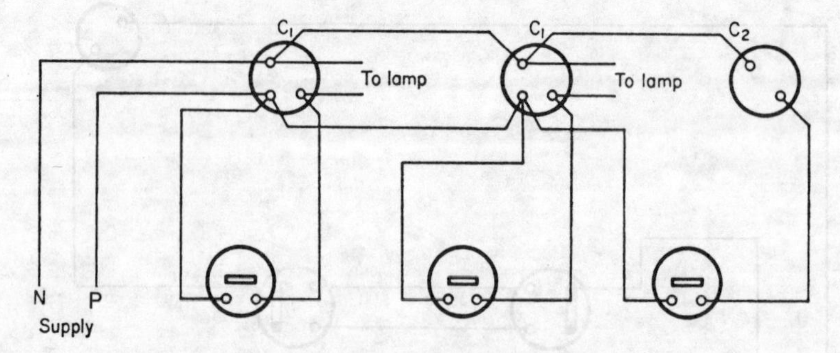

Figure 27.6. Wiring diagram of a lighting circuit using 3-plate ceiling roses and a 2-plate ceiling rose.

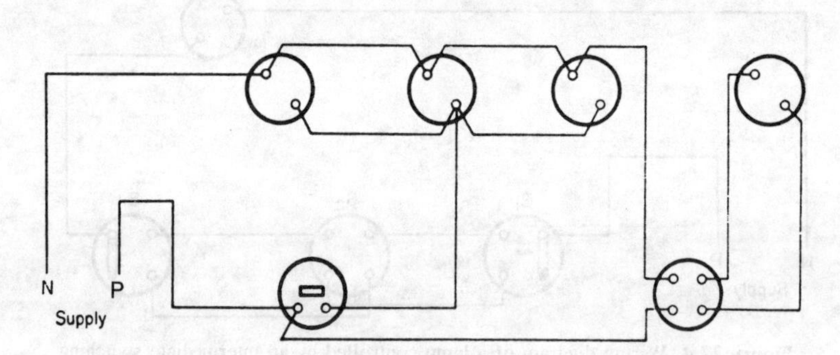

Figure 27.7. Wiring diagram of a lighting circuit with three lamps in parallel controlled by a one-way switch, and one lamp controlled by a double-pole switch.

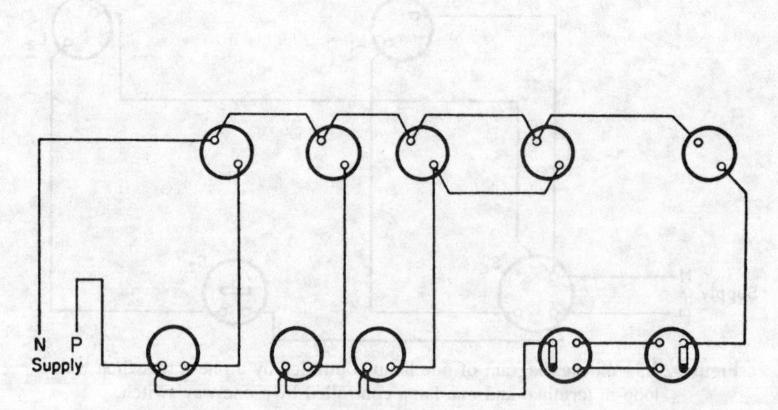

Figure 27.8. Wiring diagram of a typical lighting final circuit.

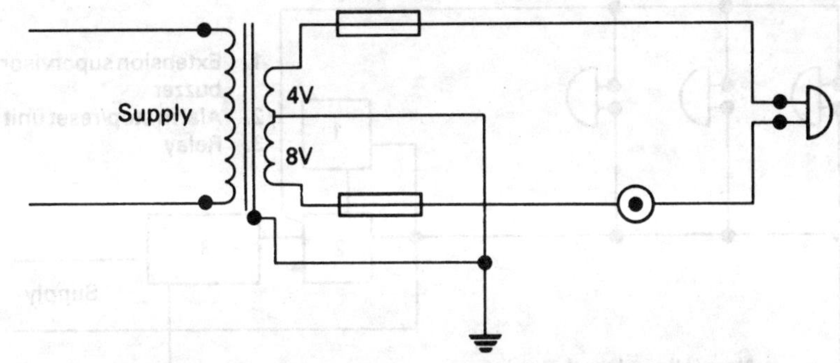

Figure 27.9. Simple single-bell circuit supplied from a bell transformer.

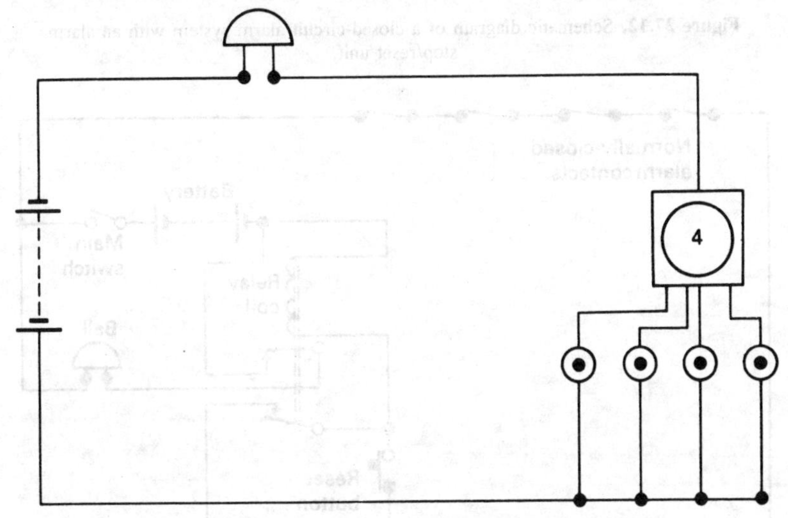

Figure 27.10. Battery-supplied bell circuit with a 4-way bell indicator. Bell operated from any bell push.

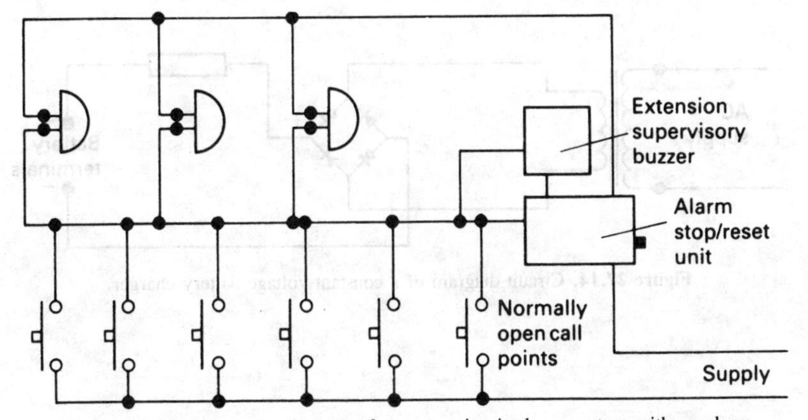

Figure 27.11. Schematic diagram of an open-circuit alarm system with an alarm stop/reset unit.

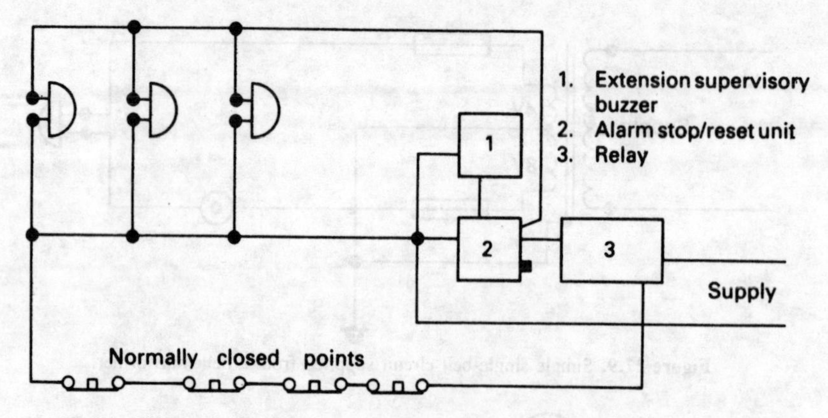

Figure 27.12. Schematic diagram of a closed-circuit alarm system with an alarm stop/reset unit.

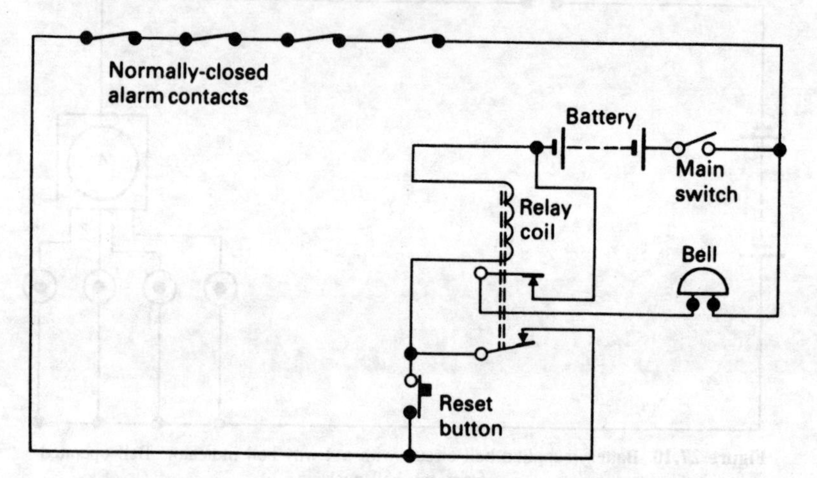

Figure 27.13. Circuit diagram of a typical closed-circuit alarm system.

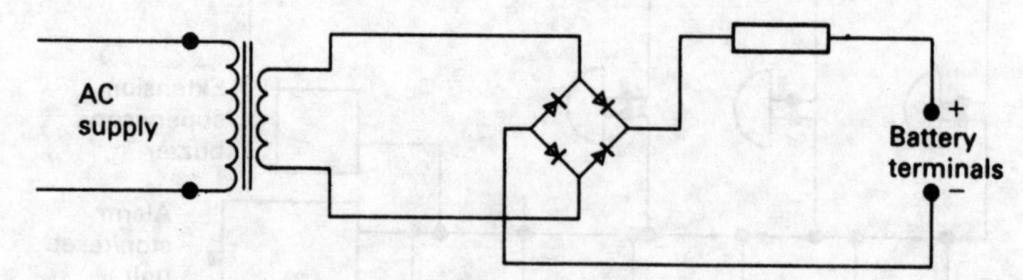

Figure 27.14. Circuit diagram of a constant-voltage battery charger.

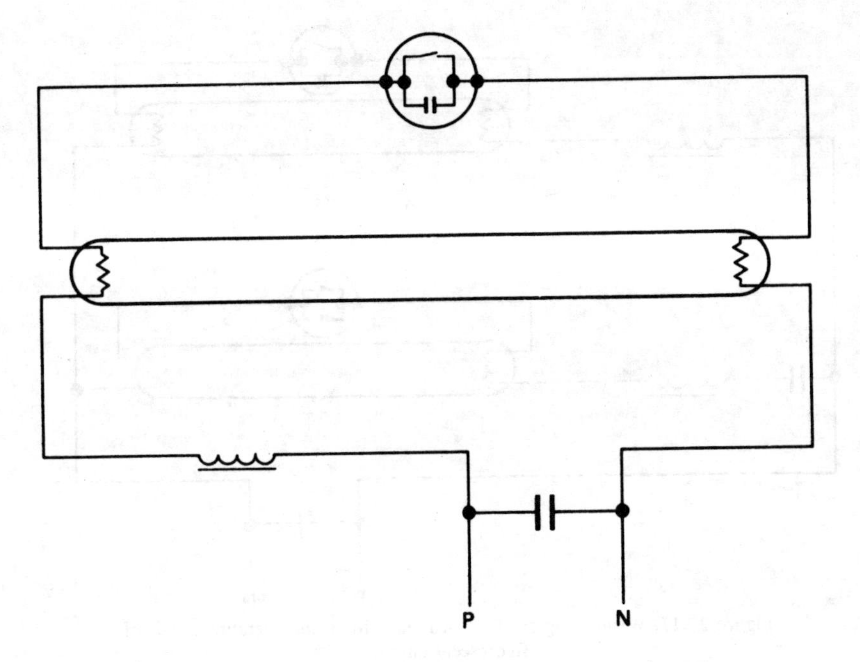

Figure 27.15. Switch-start fluorescent circuit with a glow-type starter.

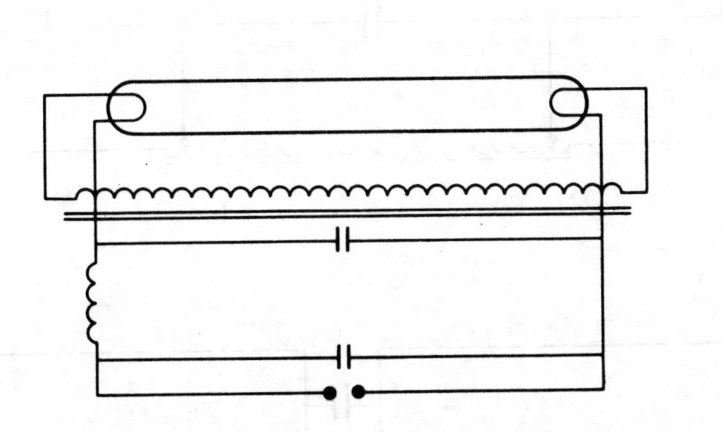

Figure 27.16. Instant-start fluorescent circuit.

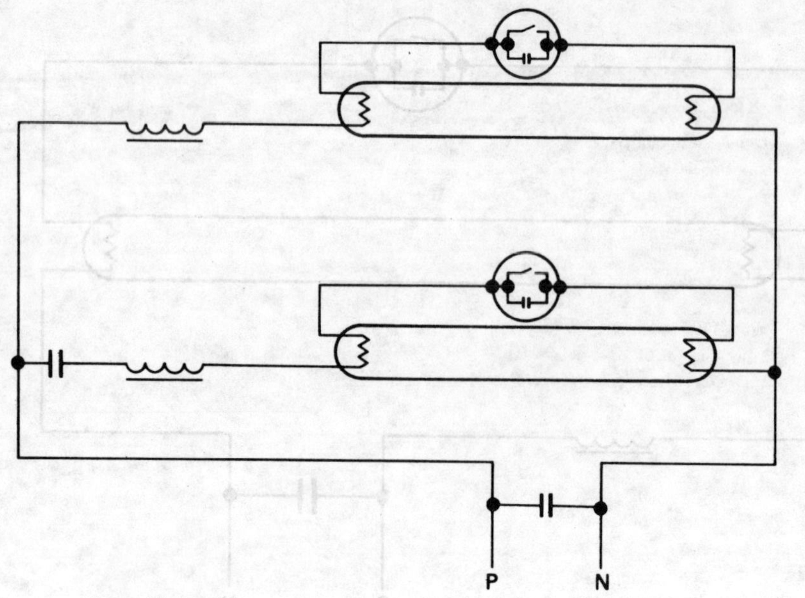

Figure 27.17. Wiring diagram for a lead-lag circuit for operating a pair of fluorescent lamps.

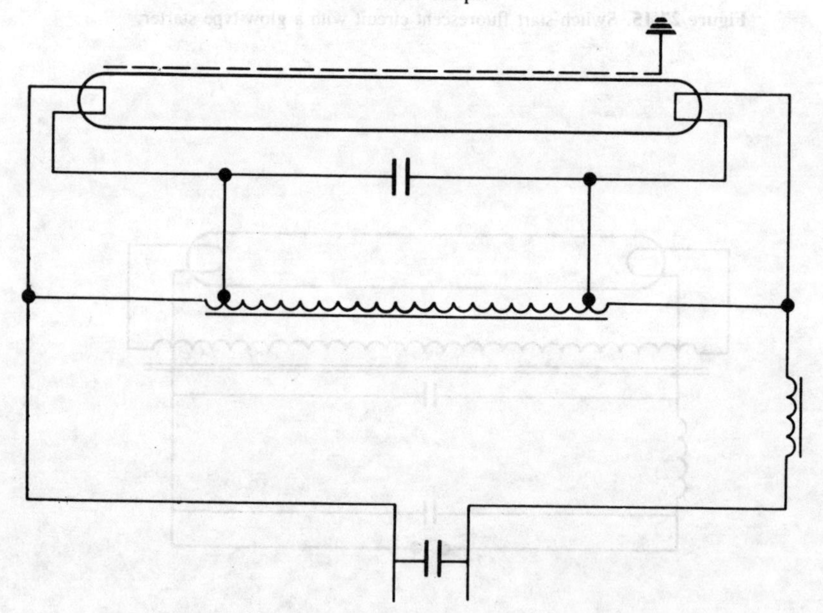

Figure 27.18. Instant-start circuit for a fluorescent lamp.

28 Fault-tracing in circuits and equipment

Types of fault

The types of fault which may occur in an electrical circuit fall into four general groups:

1. Open-circuit fault (loss of continuity).
2. Earth fault (low resistance between live conductor and earthed metalwork).
3. Short-circuit fault (low resistance between phase and neutral conductors).
4. High-value series-resistance fault (bad joint or loose connection in conducting path).

These fault types occur in lighting and power circuits, appliances, apparatus and electric motors; variations do, of course, occur with the type of electrical equipment. Before any fault can be found and rectified it is necessary for the electrician to adopt a method or system based on a sound knowledge of circuitry and electrical theory, and on experience. The electrician detailed to repair a fault circuit is in many ways like a doctor who makes his diagnosis on the basis of the symptoms revealed through a visual inspection or a test using the correct instruments. Haphazard tests carried out at random seldom lead to success in the quick location of faults. The investigation must always be based on an intelligent assessment of the fault and its probable causes, judged from its effects. In many instances, faults arise from installations or circuits which do not in some way or other comply with the requirements of the IEE Regulations, or else are used in such a manner that the abuse results in a fault. Most faults are easily located by following up reports such as 'There was a flash at the lamp'; 'The wires got red hot'; 'The lamp goes dim when it is switched on'; or 'The bedroom light will come on only when the bathroom switch is ON'. By careful questioning, these reports will enable the electrician to locate the fault quickly and restore the circuit to normal operation again. The following are some common installation defects and omissions which eventually lead to faults:

Fault-tracing in circuits and equipment

1. The provision of double-pole fusing on two-wire systems with one pole permanently earthed. This frequently occurs with final circuit distribution boards when the main and/or submain control equipment is single-pole and solid neutral.
2. Fuse protection not related to the current rating of cables to be protected. This is very often due to the equipment manufacturers fitting the fuse-carriers with a fuse-element of maximum rating for the fuse-units in the equipment.
3. Connecting boxes for sheathed-wiring systems placed in inaccessible positions in roof voids and beneath floors. Indiscriminate bunching of too many cables using screw-on or inadequate connections.
4. Insufficient protection provided for sheathed wiring, e.g. to switch positions and on joints in roof voids.
5. Incorrect use of materials, not resistant against corrosion, in damp situations (e.g. enamelled conduit and accessories and plain-steel fixing screws).
6. Inadequate or complete omission of segregation between cables and/or connections, housed within a common enclosure, supplying systems for extra-low voltage; or telecommunication and power and/or lighting operating at a voltage in excess of extra-low voltage.
7. Insufficient attention given to cleaning ends of conduit and/or providing bushings. Omission of bushings to prevent abrasion of cables at tapped entries, particularly at switch positions.

8. Insufficient precautions taken against the entry of water to duct and/or trunking systems, particularly where installed within the floor.
9. Incorrect use of PVC-insulated and/or sheathed cables and flexible cords instead of heat-resistant type, for connections to immersion heaters, thermal-storage block heaters, etc.
10. Incorrect use of braided and twisted flexible cords for bathroom pendant fittings and similar situations subject to damp or condensation.
11. The incorrect use of accessories, apparatus or appliances inappropriate for the operating conditions of the situation in which they are required to function. This often applies to agricultural and farm installations.
12. Installation of cables of insufficient capacity to carry the starting current of motors, causing excessive volt drop.
13. Incorrect rating of fuse-element to give protection to the cables supplying the motor.

Open-circuit faults

The instrument used to locate this type of fault is the continuity tester. The usual effect of this fault is that the apparatus or lamp in the circuit will not operate. The fault can be (*a*) a break in a wire; (*b*) a very loose or disconnected terminal or joint connection; (*c*) a blown fuse; (*d*) a faulty switch-blade contact. The fuse should always be investigated first. The rewireable type can be easily inspected. The cartridge type must be tested for continuity of the fuse-element. If the fuse has operated, the reason why it has done so must be found out. It is not enough to repair or replace the fuse and leave it at that. A broken wire or a disconnection will show on the continuity tester as an extremely high resistance in the kilohm or megohm ranges. Before each wire in the faulty circuit is tested in turn (live feed, switch-wire and neutral) all mechanical connections should be inspected (lampholders, junction box, plug, switch and appliance terminals). Conduit, trunking or the metal sheathing of certain wiring systems can be used as a convenient return when testing the continuity of very long conductors. In an all-insulated wiring system, other healthy conductors can be used as returns for testing purposes, making sure that the original connections are restored once the fault has been found.

Earth faults

An earth fault between a live conductor and earthed metalwork will have the same effect as a direct short-circuit: the circuit fuse will blow. To trace the fault, it is necessary to isolate the live conductor from the neutral by removing all lamps, etc., and placing all switches in the ON position. An insulation-resistant (IR) tester is used to trace this fault. Circuits should be subdivided as far as is possible to finally locate the position of the fault. The reading obtained on the instrument used will be in the low-ohms range. An earth fault on the neutral conductor seldom shows up except by an IR-to-earth test on the neutral conductor. In most instances this type of fault does not affect the operation of the circuit or the devices or equipment connected to it. However, it is important to rectify any such fault found, otherwise if it is ignored it may cause a shock and fire hazard.

Short-circuit fault

On testing the insulation resistance between the live and neutral conductors with an IR tester, the reading will show itself in the low-ohms range. Again, subdivision of the installation at the distribution board, and subdivision of the faulty circuit, is the only way to locate and confirm the position of the faults. Short circuits can occur as the result of damaged insulation, bare wire in junction boxes and fittings, or by a conductor becoming loose from terminals and moving so as to come into contact with a conductor of opposite polarity. The result of a short circuit is a blown fuse, though if there is a sufficiently high resistance in the circuit (that is, not sufficient current can flow to blow the circuit fuse) the result will be overheating of the conductors and sparking or arcing at the point of contact. The test involves the removal of all lamps and appliances from the faulty circuit, open all switches, and carry out an IR test between the live and neutral conductors. If the reading obtained is satisfactory, close each circuit switch in turn until the faulty conductor, a switch wire, is located. If a low or near-zero reading is obtained on the first test, the circuit will have to be disconnected at convenient points until the faulty wire is isolated.

High-value series-resistance faults

This type of fault is most difficult to trace as it usually means that a connection, joint or termination has become loose. The effect of this is invariably 'dim lights' or a motor 'going very slowly and heating up'. In new installations the dimness of the lamps may well be caused by a wrong connection in a junction box resulting in two or more lamps being connected in series.

Main faults in new wiring

Faults in new wiring are generally the result of careless or inadvertent wrong connections which will either blow a fuse, cause lamps to operate dimly as above, not work at all or work only when another circuit switch is placed on the ON position. If a lamp lights only when another switch in the same final circuit is ON, this indicates that the live feed to the 'faulty' lamp has been looped from the switch-wire side of the previous circuit switch instead of from the live-feed side. The fact that overloading a circuit will blow a fuse should not be overlooked.

Faults in fluorescent-lamp circuits

The following tables summarise the faults, effects and the remedies associated with fluorescent-lamp circuits.

Faults in motors and circuits

Table 28.4 summarises briefly the faults, effects and remedies associated with motors and their associated circuitry and control gear.

Faults in low-voltage circuits

Because the voltage of low- and extra-low voltage circuits is relatively small, a poor or dirty contact will immediately prevent bells and similar devices from operating. These faults are thus most difficult to trace, and it is often a matter of systematic checking for continuity (zero or near-zero) readings. The prevention of faults on ELV circuits is more often than not a matter of regular periodic maintenance attention (cleaning contacts, tightening connections, etc.) than anything else.

TABLE 28.1 Fault-finding in fluorescent lamp fittings

If a fitting fails to operate correctly, check as follows:-

Step No.	*Item*	*Tests to be applied*
1.	Supply and fuse	Check supply voltage at input to fitting. Check polarity of incoming supply and ensure frame is earthed. If fuse has blown, suspect circuit or component and find the fault before replacing fuse.
2.	Lamp	Check lamp in a good fitting and if proved faulty replace with a new lamp. Remember, never try a new lamp in a fitting which has faulty components or circuit.
3.	Circuit	Examine wiring inside the fitting and if possible check against the wiring diagram. Check insulation resistance between the circuit and the metal frame of the fitting. The resistance should be above 2 megohms. If an earth fault is found, trace the cause and replace the component.
4.	Ballast chokes	Examine for signs of overheating, if possible check continuity of windings and insulation resistance. Compare the impedance or inductance against a good replica.
5.	Capacitors	Examine for leakage or damage. If possible check the capacitance and check that the discharge resistor has a value between ¼–1 megohm. The insulation resistance between case and terminals should be above 2 megohms.
6.	Starter switches	Check operation of starter in another good circuit and, if found faulty, fit a new replacement.
7.	Ambient conditions	Remember that normal fluorescent fittings may overheat if the surrounding temperature is above 30–35 °C. Lamp starting may be difficult with some types of circuit if the temperature is below 5 °C.

TABLE 28.2 Quick-start circuits

Symptom	*Possible fault*	*Test and remedy*
Lamp fails to start — both ends glowing brightly	Wrong type of lamp or inefficient earthing	Ensure that correct grade of lamp is fitted. Quick-start lamps must have earthed metalwork within 12 mm of the tube along its full length.
	Low voltage	If supply voltage is below 220 V make sure that the leads from the choke and neutral are connected to inner terminals of the quick-start transformer unit.
	High or low temperature	Ventilate fitting if excessive temperature. Screen or enclose fitting if low temperature, or use correct grade lamp.
Lamp fails to start — one end glows brightly	Broken cathode	Test lamp in sound fitting or special test transformer unit.
	Lampholder not making contact or short-circuited	Check lampholder contacts and leads for open or short circuit. Test output voltage from quick-start secondary.
Lamp fails to start — no end glow	Open circuit or short circuit on quick-start	Test voltage applied to fittings — if correct, check circuit for open-circuited choke or quick-start or for short circuit on quick-start.
Lamp fails to start — ends glow dull and reddish	Faulty lamp	Test lamp in sound fitting or in special test transformer.
	Low cathode heating	Check output voltage from quick-start secondary.

Note: Do not make assumptions; check first the simple things, i.e. fuses and switches. Do not leave voltmeters in circuit when closing switches.

TABLE 28.3 Switch-start circuits

Symptom	*Possible fault*	*Test and remedy*
Lamp does not attempt to start — no end glow	Fuse blown	Test voltage applied to fitting — trace broken fuse. A standard Avometer may be used to obtain voltage reading. Should voltmeters be used, it may be necessary to employ a 0–400 V range for general testing, and a 0–20 or 0–25 V range for cathode heating tests. In leading power factor circuits, the voltage across the capacitor will be approximately 400 V. A test lead, with incandescent lamp of rated voltage, may be used to test for mains voltage at a fitting.
	Faulty starter	Insert starter switch in sound fitting or special test lead. A glow-switch type starter may be connected in series with a 25 W 230 V tungsten filament lamp across the main supply. The starter switch will then operate and cause the lamp to switch on and off.
	Faulty lamp	Insert tube in sound fitting or test each cathode in turn in special test transformer unit. A 12 V quick-start transformer is useful for testing cathode emission. If the 12 V winding is connected across a tube cathode, the lamp end will light up its normal colour indicating satisfactory operation. If cathode glows dull red then local ionisation is absent and indicates life-expired lamp. If one or both cathodes are broken, check for faulty circuit (short-circuit to earth or wrong control gear) before inserting new tube.
	Open circuit	Test for open circuit on choke, etc., or short to earth between choke and tube.
Lamp fails to start — bright glow from one end	Crossed leads in twin lamp fitting	Check that the correct lampholders are connected to each tube, i.e. one or both cores connected to a given tube should have the same colour.
	Short-circuit on lampholder lead	Test for short circuit across lampholder lead or for short circuit to earth on starter switch or wiring.
	Faulty lamp	Test for internal short circuits on cathode. A pen torch type of lamp tester can be used for cathode continuity reading. Alternatively a 0–30 ohm continuity tester can be used to check cathode resistance: 2–10 ohms (cold).
Lamp does not attempt to start — both ends glow brightly	Short-circuit on starter switch or associated wiring	Test starter switch in sound fitting or special test lead. If satisfactory, test switch socket and wiring for short circuit.
Lamp flashes on and off — fails to maintain discharge	Faulty lamp	Test lamp in sound fitting or special test transformer. At end of life other symptoms are reduced light output, increased flicker and reddish glow from cathodes.
	Low voltage or incorrect tapping	Test voltage at terminal block of fitting — if low, check external wiring for excessive drop (fuse holders, etc.). Check choke tapping against 'sustained' voltage at terminals. If voltage is persistently low reduce tapping or convert to series capacitor circuit.
	Faulty starter	Test starter switch in sound fitting or special test lead.
	Low temperature	Screen open type fittings or use low-temperature grade lamps.
	Crossed leads in twin fitting	Check that the correct lampholders are connected to each lamp, i.e. one or both cores connected to a given tube should have the same colour.

TABLE 28.4 Motors and motor circuits

Fault	*Possible cause*	*Corrective action*
1. Vibration	Uneven foundations	Check level and alignment of base and realign.
	Defective rotor	See (8).
	Unbalance	Uncouple from driven machine, remove motor pulley or coupling. Run motor between each of these operations to determine whether unbalance is in the driven machine, pulley, or rotor. Rebalance.
2. Frame heating*	Excessive load	See (5).
	Foreign matter in airgap or cooling circuit	Check airgap, dismantle motor and clean.
	Excessive ambient temperature	Motors supplied to BS 2613 : 1957 are intended for operation in an ambient not exceeding 40 °C. Where the ambient exceeds 40 °C a motor of corresponding lower temperature rise should be used.
	Partial short or open circuit in windings	Check windings with suitable meter. If defective, repair or return to manufacturer.
3. Bearing heating	Too much grease	Remove surplus grease.
	Too little grease	Wash bearings and replensih with grease.
	Incorrect assembly	Ensure bearing assembled squarely on shaft.
	Bearing overloads	This may be due to misalignment of the drive, excessive end thrust imposed on motor, or too great a belt tension. Take appropriate steps to reduce the load on the bearing.
4. Brushes heating	Excessive load	See (5).
	Brushes not bedding or sticking in holders	Carefully re-bed or clean brushes and adjust pressure.
	Brush chatter	Ensure that commutator is true without high or low bars and adjust brush pressure.
	Incorrect grade of brushes	Ensure that brushes are those specified by the motor manufacturer.
5. No rotation	Supply failure, either complete or single phase	Disconnect motor immediately — with a single-phase fault serious overloading and burn-out may rapidly occur. Ensure that correct supply is restored to motor terminals.
	Insufficient torque	Check starting torque required and compare with motor rating, taking into account type of starter in use. Change to larger motor or to different type of starter.
	Reversed phase	Check connections in turn and correct.
6. Steady electrical hum	Running single phase	Check that all supply lines are live with balanced voltage.
	Excessive load	Compare line current with that given on motor nameplate. Reduce load or change to larger motor.
	Reversed phase	Check connections in turn and correct.
	Uneven airgap	Check with feelers. If due to worn bearings fit new ones.
7. Pulsating electrical hum	Defective rotor	Check speed at full load. If it is low and if there is a period swing of current when running, a defective rotor is indicated, and the matter should be referred to the manufacturers.
	Defective wound rotor. Loose connection, partial short-circuit, etc.	On a wound rotor machine check should be made of rotor resistance and open-circuit voltage between slip-rings.
8. Mechanical noise	Foreign matter in airgap	Check airgap, dismantle rotor and clean.
	Bearings damaged	Check with a listening stick. If confirmed try rotating outer race of bearing 180°. If still unsatisfactory fit new bearing.
	Couplings out of line	Check coupling gap and realign.

* The frame temperature should be checked with a thermometer; the reading so obtained will be approximately 10 °C lower than the actual temperature of the windings.

Appliances and apparatus

The consumer-user range of appliances (cleaners, irons, kettles, fires and so on) have generally to withstand much handling which eventually leads to faults. One of the most common is a break in the continuity of a conductor of three-core flexible cord, particularly where the cord emerges from the plug or enters the appliance connector, or the appliance cable-entry position. This break in the conductor is most often the result of excessive movement of the cord at one point over a long time. The fault is usually identified by the intermittent working of the appliance and perhaps signs or smells, of burning. Conductors are often pulled away from their terminals to cause short circuits or intermittent high series-resistance contacts.

TABLE [illegible] Motors and motor circuits

Fault	Possible cause	Corrective action
1. Vibration	Uneven foundations Defective rotor Unbalanced [illegible]	Check level and alignment of base and fixings. See (3). Uncouple from driven machine, remove motor pulley or coupling. Run motor between each of these operations to determine whether imbalance is in the driven machine, pulley or rotor. Rebalance.
2. Frame heating	Excessive load Foreign matter obstructing cooling circuit Excessive ambient temperature Partial short or open circuit in windings	See (5). Check airflow, dismantle motor and clean. Motors supplied to BS 2613 are intended for operation in an ambient not exceeding 40 °C. Where the ambient exceeds 40 °C a motor of corresponding lower temperature rise should be used. Check windings with suitable meter. If defective, repair or return to manufacturer.
3. Bearing heating	Too much grease Too little grease Incorrect assembly [illegible] overloads	Remove surplus grease. Wash bearings and replenish with grease. Ensure bearing assembled squarely on shaft. This may be due to misalignment of the drive, excessive end thrust imposed by coupling or too great a belt tension. Take appropriate steps to reduce the load on the bearings.
4. Brush sparking	Excessive load Brushes not bedding or sticking in holders Brush pressure Incorrect grade of brushes	See (5). Carefully re-bed or clean brushes and adjust pressure. Ensure that commutator is true without high or low bars and adjust brush pressure. Ensure that brushes are those specified by the motor manufacturer.
5. [illegible]	Supply failure, either complete or single phase Insufficient torque	Disconnect motor immediately — with a single-phase fault serious overheating and burnout may rapidly occur. Ensure that correct supply is restored to motor terminals. Check starting torque required and compare with motor rating, taking into account type of starter in use. Change to larger motor or to different type of starter.
	Reversed phase	[illegible] connections in turn and correct.
6. [illegible]	Running single phase Excessive load	Check that all supply line fuses are intact with balanced voltage. Compare line current with that given on motor nameplate. Reduce load or change to larger motor.
	Reversed phases [illegible]	Check connections in turn and correct. [illegible]
7. [illegible]	Defective rotor	Check speed at full load. If it is low and there is a periodic swing of current when running, a defective rotor is indicated, and the matter should be referred to the manufacturers.
	Defective wound rotor, loose connection, partial short circuit	On a wound rotor machine check should be made of rotor resistance and open-circuit voltage between slip rings.
8. Mechanical noise	Foreign matter in air gap Bearings damaged	Check air gap, dismantle rotor and clean. Check for rumbling noise. If confirmed try rotating outer race of bearing 180°. If still unsatisfactory fit new bearing.
	Coupling loose	Check coupling grip and realign.

* If the temperature is measured with a thermometer the reading so obtained will be approximately 10 °C lower than the actual temperature of the windings.

Appliances and apparatus

The consumer uses a range of appliances (cleaners, irons, kettles, fires and so on) that generally undergo much handling which eventually leads to faults. One of the most common is a break in the continuity of a conductor of three-core flexible cord, particularly where the cord emerges from the plug or enters the appliance connector, or the appliance cable-entry position. This break in the conductor is most often the result of excessive movement of the cord at one point over a long time. The fault is usually identified by the intermittent working of the appliance and perhaps signs of 'flashes' of burning. Conductors are often pulled away from their terminals to cause short circuits or, more often, high series-resistance contacts.

Part B

Electrical workshop practice

29 Workshop safety

Safety in an electrical workshop means much more than the provision which enables persons to use electricity without risk of shock. Means to prevent accidents, not necessarily of an electrical nature, must at all times be part and parcel of workshop procedure and facilities. There are two main causes of accidents in factories and workshops. First, carelessness. When any task is done often enough it becomes familiar to the point at which boredom sets in; when this happens, original high standards begin to slide. The result is often an accident to that person or to an innocent fellow-worker, which may or may not be fatal. Secondly, there is the ignorance factor. This does not mean lack of intelligence. Rather it means a lack of sufficient technical knowledge to perform a task efficiently and to leave the task in a condition which is safe. Many tasks seem simple when done by the expert. If there is a lack of knowledge, a task should be done under expert supervision.

Electric shock being one of the main occupational hazards of the electrician, the following points should be observed:

1. Never take it for granted that a circuit is 'dead'. Always check thoroughly using a test lamp or a neon tester. Never check for 'live' by touching conductors with fingertips.
2. If a circuit must be worked on while 'live', it is always best to leave the work to a more experienced electrician. Otherwise, use adequately insulated tools, rubber gloves, and stand on rubber matting. The higher the voltage the greater the care needed.
3. If working on a circuit to which other persons may have access always remove the circuit fuses and indicate that the circuit has been made 'dead' deliberately. Keep the fuses in your pocket until the job is finished.

So far as general care in the workshop is concerned, always be careful when working on or beside moving machinery. Goggles must be worn when grinding metal. All machinery should have guards. Ladders should be placed at reasonable angles to the vertical and be placed on a non-slip surface.

First aid

Though first aid is meant only to be a temporary measure until skilled assistance can be obtained, it is no less important to have some basic knowledge of the treatment of injuries. Every factory and workshop must have a first-aid box; this is a legal requirement.

Treatment for electric shock

There are two methods used to treat electric shock. In any case, speed is essential as even a few minutes' delay may cause death. A severe electric shock will affect the nerves which control the breathing and the action of the heart. In the treatment, effort is made to get the patient's body working normally, even while unconscious.

Before the treatment is begun, it is important to make sure that the patient is not still in contact with the live object. If this is the case, then the patient must not be moved with bare hands, but should be pulled away from the live contact using a jacket, a chair, dry folded newspaper, a belt, rope or a length of dry wood. Standing on newspapers, dry wood, or dry clothes also helps to increase the insulation factor between the rescuer and the floor which may be conducting. Any obstructions to the patient's breathing should be removed (e.g. tight collar). Artificial respiration should be given immediately the patient is clear from the live electrical contact.

Holger Nielsen method. Place the patient face downwards. See that the forehead is resting on the hands (placed one over the other) so that the nose and mouth are clear. A rolled-up jacket will help to raise the head. If the patient's tongue has been swallowed, two or three firm slaps with the flat of the hand on the area between the shoulders will bring it forward.

The rescuer should now kneel in front of the patient, a knee beside the right of the head and a foot to the left at the patient's elbow. The arms should slope forward so that the hands lie close together on the patient's shoulders; the wrists should be over the top of the shoulder-blades. Begin the movements as follows:

1. Move forward and press down with a light pressure to drive air from the lungs. This movement should last about two seconds (count: ONE, TWO).
2. Slide hands quickly down to patient's elbows, this movement taking about a second (THREE).
3. Raise the elbows slightly. The rescuer's body should move backwards a little. This movement induces air into the patient's lungs and should take about two seconds (FOUR, FIVE).
4. Lower elbows and slide hands to patient's back to resume original position (SIX).

The above movements should be repeated until the patient recovers. The movement allows for about nine respirations per minute. It is recommended that efforts to save life should be maintained for about four hours. Patients who have recovered should always be seen by a doctor, who should be sent for as soon as the accident has been discovered.

Mouth-to-mouth method. This method is sometimes known as the 'kiss-of-life' method of shock-treatment, and has proved easy to apply and as effective as the Holger Nielsen method.

1. Place the patient on his back and sit or kneel by the side of his head. The head should be held in both hands. One hand should press upwards; the other pushing the jaw upwards and forwards. This position is to ensure that the patient's lungs get maximum air to them.
2. Close the patient's nose with one hand.
3. Seal your lips round the patient's mouth. Blow air steadily into the patient's mouth until the swelling lungs cause the chest to rise.
4. Remove your mouth. Turn your head aside and take a deep breath.

Repeat this cycle six times as quickly as possible. Then continue at about 10-second intervals.

As a general precaution, a thin handkerchief may be placed over the patient's mouth or nose if desired. If the patient's lungs seem to be obstructed, this could well be the tongue which may have slipped back. The patient's head should always be kept pressed back.

Electric burns

These are best seen by a first-aid expert or by a doctor. Burnt clothing should not be removed. And blisters should not be broken. Sterilised dressing may be applied if thought necessary. Patient may be given water, tea, coffee or some other liquid, but no alcohol. Keep the patient warm at all times.

30 Electrical materials 1: Conductors

In electrical work, materials used fall into five general groups: conductors, semiconductors, insulators, magnetic and constructional. Each group finds its place in the equipment, circuitry and devices to be found in all aspects of the electrical engineering field. Some materials are used for different purposes. Iron, for instance, is a magnetic material and is also used for constructional purposes: and with other elements to form a ferrous alloy, it is a resistive conductor as found in heating elements. Chapter 31 deals with insulating materials.

Conductors

In electrical work, a 'conductor' means a material which will allow the free passage of an electric current along it, and which presents negligible resistance to the current. If the conducting material has an extremely low resistance (e.g. a copper cable) there will, normally, be no effect when the conductor carries a current. If the conducting material has a significant resistance (e.g. iron wire) then the conductor will show the effects of an electric current passing through it, usually in the form of a rise in temperature to produce a heating effect. It should be remembered that the conduction of electric currents is offered not only by metals, but by liquids (e.g. water) and gases (e.g. neon). Conductors by nature differ so enormously from insulators in their degree of conduction that the materials which offer high resistance to an electric current are classed as insulators. Those materials which fall in between the two are classed as semiconductors (e.g. germanium).

Copper

This metal has been known to man since the beginning of recorded history. Copper was connected with the earliest electrical effects such as, for instance, that made by Galvani in 1786 when he noticed the curious behaviour of frogs' legs hung by means of a copper hook from an iron railing (note here the two dissimilar metals). Gradually copper became known as an electrical material; its low resistance established it as a conductor. One of the first applications of copper as a conductor was for the purpose of signalling; afterwards the commercial generation of electricity looked to copper for electrical distribution. It has thus a prominent place and indeed is the first metal to come to mind when an electrical material is mentioned. As a point of interest, the stranded cable as we know it today has an ancient forebear. Among several examples, a bronze cable was found in Pompeii (destroyed AD 79); it consisted of three cables, each composed of fifteen bronze wires twisted round each other.

Copper is a tough, slow-tarnishing and easily worked metal. Its high electrical conductivity marks it out for an almost exclusive use for wires and cables, contacts, and terminations. Copper for electrical purposes has a high degree of purity, at least 99.9 per cent. This degree of purity results in a conductivity value only slightly less than that of silver (106 to 100). As with all other pure metals, the electrical resistance of copper varies with temperature. Thus, when there is a rise in temperature, the resistance also increases. Copper is available as wire, bar, rod, tube, strip and plate. Copper is a soft metal; to strengthen it certain elements are added. For overhead lines, for instance, copper is required to have a high-tensile strength and is thus mixed with cadmium. Copper is also reinforced by making it surround a steel core, either solid or stranded.

Copper is the basis of many of the cuprous alloys found in electrical work. Bronze is an alloy of copper and tin. It is fairly hard and can be machined easily. When the bronze contains phosphorus, it is known as phosphor-bronze, which is used for spiral springs. Gunmetal (copper, tin and zinc) is used for terminals.

Copper and zinc become brass which is familiar as terminals, cable legs, screws and so on, where good conductivity is required coupled with resistance to wear. Copper oxidises slowly at ordinary temperatures, but rapidly at high temperatures; the oxide skin is not closely adherent and can be removed easily.

Aluminium
The use of aluminium in the electrical industry dates back to about the turn of this century when it was used for overhead-line conductors. But because in the early days no precautions were taken to prevent the corrosion which occurs with bimetallic junctions (e.g. copper cable to aluminium busbar) much trouble was experienced which discouraged the use of the metal. Generally speaking, aluminium and its alloys are used today for electrical purposes because of (*a*) weight; (*b*) resistance to corrosion; (*c*) economics (cheaper than copper); (*d*) ease of fabrication; (*e*) non-magnetic properties. Electrical applications include cable conductors, busbars, castings in switchgear, and cladding for switches. The conductor bars used in the rotor of squirrel-cage-induction ac motors are also of aluminium on account of the reduced weight afforded by the metal. Cable sheaths are available in aluminium. When used as conductors, the metal is either solid or stranded.

An oxide film is formed on the metal when exposed to the oxygen in the atmosphere. This film takes on the characteristics of an insulator, and is hard enough to withstand some considerable abrasion. The film also increases the corrosion-resisting properties of aluminium. Because of this film it is important to ensure that all electrical contacts made with the metal are initially free from it; if it does form on surfaces to be mated, the film must be removed or broken before a good electrical contact can be made in a joint. Because the resistivity of aluminium is greater than that of copper, the cross-sectional area of the conductor for a given current-carrying capacity must be greater than that for a copper conductor.

Zinc
This metal is used mainly as a protective coating for steel and may be applied to the steel by either galvanising, sherardising or spraying. In electrical work it is found on switchgear components, conduit and fittings, resistance grids, channels, lighting fittings and wall brackets. Galvanising is done by dipping iron or steel objects into molten metal after fluxing. Mixed with copper, the zinc forms the alloy brass. Sherardising is done by heating the steel or iron object to a certain temperature in zinc dust, to result in an amalgamation of the two metals, to form a zinc-iron alloy.

Lead
Lead is one of the oldest metals known to man. Lead is highly resistant to corrosion. So far as the electrical application of lead is concerned, apart from its use in primary and secondary cells, cable sheathing in lead was suggested as early as 1830–45. This period saw the quantity production of electrical conductors for inland telegraphs, and thoughts turned to the possibility of prolonging the life of the conductors: the earliest suggestions were that this could be done by encasing them in lead. Today lead is used extensively. Lead is not used pure; it is alloyed with such metals as tin, cadmium, antimony and copper. Its disadvantage is that it is very heavy; it is also soft, even though it is used to give insulated cables a degree of protection from mechanical damage. One of its principal properties is its resistance to the corrosive effects of water and acids. It has a low melting point; this fact is made use of in the production of solder, where it is alloyed with tin for cable-jointing work. Lead alloyed with tin and copper is used as white metal for machine bearings.

Nickel
This metal is used in conjunction with iron and chromium to form what is known as the resistive conductors used as heating elements for domestic and industrial heating appliances and equipment. The alloy stands up well to the effects of oxidation. Used with chromium only the alloy is non-magnetic; with iron it is slightly magnetic. It has a high electrical resistivity and low temperature coefficient. The most common alloy names are Nichrome and Brightray and Pyromic. Pure

nickel is found in wire and strip forms for wire leads in lamps, and woven resistance mats, where resistance to corrosion is essential.

Carbon

This material is used for motor brushes (slip-ring and commutator), resistors in radio work. It has a negative temperature characteristic in that its resistance decreases with an increase in temperature.

Ferrous metals

These metals are based on iron and used for the construction of many pieces of equipment found in the electrical field (switches, conduit, cable armouring, motor field-poles and so on). Because iron is a magnetic material, it is used where the magnetic effect of an electrical current is applied to perform some function (e.g. in an electric bell).

The choice of magnetic materials today is extremely wide. For practical purposes magnetic materials fall into two main classes: permanent (or hard) and temporary (or soft). Permanent magnetic materials include tungsten and chromium steel and cobalt steel: when magnetised they retain their magnetic properties for a long time. Cobalt-steel magnets are used for measuring instruments, telephone apparatus and small synchronous motors. Soft magnetic materials do not retain their magnetism for any appreciable time after the magnetising force has been withdrawn. In a laminated sheet form they are found in transformer cores and in machine poles and armatures and rotors. Silicon–iron is the most widely used material for cores.

Rare and precious metals

In general, precious metals are used either for thermocouples or contacts. Among the metals used are silver, gold, platinum, palladium and iridium. Sometimes they are used as pure metals, otherwise as an alloy within the above group or with iron and copper, where special characteristics are required. For instance, a silver–iron alloy contact has a good resistance to sticking and is used in circuits which are closed with a high inrush (e.g. magnetising currents associated with inductors, electromagnets and transformers). It is used also for small motor-starter contacts. The alloy maintains low contact resistance for very long periods. The following are some applications of rare and precious metals in contacts:

Circuit-breakers. Silver, silver–nickel, silver–tungsten.

Contactors. Silver, silver–tungsten.

Relays. Silver, platinum, silver–nickel.

Starters. Platinum, rhodium, silver, coin silver. Silver is used for the fuse-element in HRC fuses.

Mercury. This material is used almost exclusively for mercury switches. In a vapour form it is used in fluorescent lamps (low-pressure lamps) and in the high-pressure mercury-vapour lamp.

Semiconductors

Oxides of nickel, copper, iron, zinc and magnesium have high values of resistivity; they are neither conductors nor insulators, and are called semiconductors. Other examples are silicon and germanium. When treated in certain ways, these materials have the property of being able to pass a large current in one direction while restricting the flow of current to a negligible value in the other direction. The most important application for these materials is in the construction of rectifiers and transistors.

Conducting liquids

Among the liquids used to conduct electric currents are those used as electrolytes: sulphuric acid (lead-acid cells); sal ammoniac (Leclanché cells); copper sulphate (in simple cells); caustic potash (nickel–cadmium cells). When salts are introduced to water the liquid is used as a resistor.

Conducting gases

In electrical work, so far as the practical electrician is concerned, the conducting gases are those used for electric-discharge lamps: neon, mercury vapour, sodium vapour, helium.

31 Electrical materials 2: Insulators

An insulator is defined as a material which offers an extremely high resistance to the passage of an electric curent. Were it not for this property of some materials we would not be able to apply electrical energy to so many uses today. Some materials are better insulators than others. The resistivity of all insulating materials decreases with an increase in temperature. Because of this, a limit in the rise in temperature is imposed in the applications of insulating materials, otherwise the insulation would break down to cause a short circuit or leakage current to earth. The materials used for insulation purposes in electrical work are extremely varied and are of a most diverse nature. Because no single insulating material can be used extensively, different materials are combined to give the required properties of mechanical strength, adaptability and reliability. Solids, liquids and gases are to be found used as insulation.

Insulating materials are grouped into classes:

Class A — Cotton, silk, paper and similar organic materials; impregnated or immersed in oil.

Class B — Mica, asbestos, and similar inorganic materials, generally found in a built-up form combined with cement binding cement. Also polyester enamel covering and glass-cloth and micanite.

Class C — Mica, porcelain glass quartz and similar materials.

Class E — Polyvinyl acetal resin.

Class H — Silicon–glass.

The following are some brief descriptions of some of the insulating materials more commonly found in electrical work.

Rubber

Used mainly for cable insulation. Cannot be used for high temperatures as it hardens. Generally used with sulphur (vulcanised rubber) and china clay. Has high insulation-resistance value.

Polyvinyl chloride (PVC)

This is a plastics material which will tend to flow when used in high temperatures. Has a lower insulation-resistance value than rubber. Used for cable insulation and sheathing against mechanical damage.

Paper

Must be used in an impregnated form (resin or oil). Used for cable insulation. Impregnated with paraffin wax, paper is used for making capacitors. Different types are available: kraft, cotton, tissue, pressboard.

Glass

Used for insulators (overhead lines). In glass fibre form it is used for cable insulation where high temperatures are present, or where areas are designated 'hazardous'. Requires a suitable impregnation (with silicone varnish) to fill the spaces between the glass fibres.

Mica

This material is used between the segments of commutators of dc machines, and under slip-rings of ac machines. Used where high temperatures are involved such as the heating elements of electric irons. It is a mineral which is present in most granite-rock formations; generally produced in sheet and block form. Micanite is the name given to the large sheets built up from small mica splittings and can be found backed with paper, cotton fabric, silk or glass-cloth or varnishes. Forms include tubes and washers.

Ceramics

Used for overhead-line insulators and switchgear and transformer bushings as lead-ins for cables and conductors. Also found as switch-bases, and insulating beads for high-temperature insulation applications.

Bakelite
A very common synthetic material found in many aspects of electrical work (e.g. lampholders, junction boxes), and used as a construction material for enclosing switches to be used with all-insulated wiring systems.

Insulating oil
This is a mineral oil used in transformers, and in oil-filled circuit-breakers where the arc, drawn out when the contacts separate, is quenched by the oil. It is used to impregnate wood, paper and press-board. This oil breaks down when moisture is present.

Epoxide resin
This material is used extensively for 'potting' or encapsulating electronic items. In larger castings it is found as insulating bushings for switchgear and transformers.

Textiles
This group of insulating materials includes both natural (silk, cotton, and jute) and synthetic (nylon, Terylene). They are often found in tape form, for winding-wire coil insulation.

Gases
Air is the most important gas used for insulating purposes. Under certain conditions (humidity and dampness) it will break down. Nitrogen and hydrogen are used in electrical transformers and machines as both insulants and coolants.

Liquids
Mineral oil is the most common insulant in liquid form. Others include carbon tetrachloride, silicone fluids and varnishes. Semi-liquid materials include waxes, bitumens and some synthetic resins. Carbon tetrachloride is found as an arc-quencher in high-voltage cartridge type fuses on overhead lines. Silicone fluids are used in transformers and as dashpot damping liquids. Varnishes are used for thin insulation covering for winding wires in electromagnets. Waxes are generally used for impregnating capacitors and fibres where the operating temperatures are not high. Bitumens are used for filling cable-boxes; some are used in a paint form. Resins of a synthetic nature form the basis of the materials known as 'plastics' (polyethylene, polyvinyl chloride, melamine and polystyrene). Natural resins are used in varnishes, and as bonding media for mica and paper sheets hot-pressed to make boards.

32 Workshop measurements

Many kinds of measurement are taken in everyday work in the workshop: lengths, widths, angles, thicknesses and so on. For some measurements simple tools are available, such as the rule. For others, particularly for accurate work, tools are more complicated (e.g. the micrometer). Each measuring instrument is designed to do a certain job: to provide information in recognisable units for comparison with a group or set of similar units relating to the shape and size of a piece of work. Electrical measurements are also made to discover the electrical properties of materials and the circuits in which they are incorporated. This chapter is no more than a brief summary of the instruments and tools used for measurement work in the workshop.

'Line' and 'end' measurement

It is one thing to check the accuracy of one's work; it is another thing to indicate its size by actual measurement. A length may be expressed as the distance between two lines (called 'line measurement'), or as the distance between two faces (called 'end measurement'). The most common example of line measurement is that which uses the rule, which has divisions shown as lines marked on its surface. For end measurement we use calipers, the micrometer, solid length bars and so on to obtain size.

The rule

For workshop measurement work it is best to buy a good rule of the engineer's type. For work outside the workshop, the electrician uses a boxwood rule, because he does not require generally the same degree of accuracy. The most useful and convenient markings on the engineer's rule are inches on one face and metric units on the other. A good rule is worth looking after. It should not be used as a feeler gauge, screwdriver blade or other purpose which may mar its ability to do its specific job. In particular its ends should be protected from wear, because they form one end of a dimension. A rounded corner end could well introduce an inaccuracy of 0.25 or 0.5 mm. Rusting of the rule can be prevented by keeping it in an oiled cloth or rustproof paper. When taking measurements with a rule it is best to hold it so that the gradation lines are as near as possible to the faces to be measured. Some rules are provided with bevelled edges, so that 'parallax' is reduced to the minimum. This term is used for the type of error which results from looking at, for example, a clock at an acute angle instead of from the front of it.

Calipers

To measure the diameter of a circular part involves straddling across it to obtain the length of its greatest dimension. The rule is not always the right tool for this, so we use the calipers. The shape of calipers varies according to whether they are to be used for external or internal work. They may be stiff-jointed at the hinge of the legs (opening is maintained by the friction at the joint); or else the joint may be free and spring-controlled (opening is adjusted and maintained by a nut working on a screw). Screw-controlled spring calipers are more easily adjusted. When the calipers are adjusted and indicate the dimension between their legs, the dimension should be read off either on a rule or micrometer. Calipers should be used with care in the workshop and not knocked around so that they become damaged.

The micrometer

When a part has to be measured to, say, the third place after the decimal point in the Imperial system of units, or the second place in the metric system, a more accurate method of measurement is needed than can be obtained with the use of the rule. The micrometer is used here. It consists of a semi-circular frame with a cylindrical extension (the barrel) at its right end, and a hardened anvil inside, at the left end. The bore of the barrel is screwed with a very fine thread and the spindle, which is attached to a thimble, screws through. The barrel is gradated in divisions of a unit, as is the rim of the thimble. Measurement is taken between the face of the anvil and the end of the spindle. The accuracy of the micrometer depends on the price paid for it. It should be treated with care as it is a most valuable and useful instrument.

The vernier calipers

This measuring device gives an 'end' measurement. The positions of the jaws of the instrument are, however, controlled from a 'line' scale. Transfer of one measurement to the other is made possible by the vernier scale. A vernier scale is the name given to any scale which makes use of the difference between two scales which are nearly, but not quite, alike, so that small differences can be obtained. The vernier is made in various sizes from 150 mm upwards. It is not used for straddling round bars in the same way as a micrometer, but may be used for measuring large diameters on their ends, or large bores.

Gauges

While an instrument is used to measure a dimension, the gauge is used to check the accuracy of a piece of work, without any particular reference to its size. There is an extremely wide range of gauges available, all of which have a particular function in the workshop. Included in the range are *hole gauges, limit plug gauges* and *plate gauges*. The material for gauges is extremely hard and resistant to wear. For general electrial work, the wire gauge is used to check the size of a wire. Other gauges in general use are often associated with a manufacturer's product. For example, to check the size of MICS cable, some manufacturers issue a gauge with a large vee-slot and marked off in the actual size of the cable. Where important contracting jobs are being carried out to rigid specifications, it is sometimes necessary to check off, say, the thickness of conduit walls.

Electrical measurements

The measurement of electrical properties of materials and electrical quantities is made by using instruments, the basic principles of which are discussed elsewhere in this book. The properties usually measured are insulation resistance, continuity and conductor resistance. The electrical quantities are current and voltage, generally associated with a circuit in operation (e.g. an electric iron on test). Chapter 17 of this book discusses electrical measurements. Many of the most accurate instruments, particularly the multi-range types, are expensive to buy. But by the time the years of apprenticeship are out, it should be possible for the newly-fledged journeyman with a respect for his work to make the purchase of a small multi-range instrument which, even if the ranges are limited, will still form an extremely useful part of his tool kit. The more expensive instruments are supplied by the employer and their use is generally restricted to experienced men who are able, not only to use them correctly, but to interpret their readings.

33 Workshop practice

Basic workshop operations. A great deal of the activity of the electrician or electrical fitter in the electrical workshop is closely associated with mechanical engineering craft practice. For instance, metals are cut, drilled, tapped and so on. In the context of maintenance and repairs on electrical machines and apparatus, the element of mechanical engineering is often greater than that of the electrical side of things. This chapter is no more than a reasonably comprehensive summary of the techniques involved in some of the more common activities usually found in an electrical workshop. Some of the activities will also be those done by the electrician outside the workshop — on site, in a customer's home and in places where workshop facilities are not available. But, in general, the techniques remain the same.

Riveting

When two flat surfaces (e.g. plates) have to be fastened together to form a permanent joint, riveting is a satisfactory method. Rivets are classed according to the shape of the head. The round snap-head is the most commonly used. But if the projecting head is an inconvenience, the countersunk type enables a flush-head finish to be obtained — though it does not give such an efficient joint. Rivets put in and riveted-up hot give the best results, particularly for heavy work. However, for light work, copper, brass and aluminium rivets may be used cold. For very light work, bifurcated and hollow rivets may be obtained. But these do not give such a satisfactory joint as the solid rivet. Riveted joints to plates or flat bar may be made either by lapping over the edges of the plates and fastening with one or two rows of rivets (lap joint). Or else, the edge of the plates may be butted together and the joint completed by holding them together with one or two cover straps and riveting (butt joints).

When preparing the plates for the joint they should, if possible, be clamped together with the top plate marked out for the holes. The holes are then drilled in all the plates at once. Using this method, there will thus be no doubt about all the holes being in exact alignment when the rivets are put in them. If one plate has already been drilled, it should be clamped in position and the holes marked through for drilling in the other plates; or the holes can be used as guides for the drill itself. The riveting process, either using heat or while cold, is the closing of the rivet: supporting the rivet head while the plain end is riveted over using an appropriate cupped punch or the ball-end of a hammer. The process is to swell the metal in the hole to fill it and so pull the plates tightly together. Care should be taken to see that the rivet end is spread evenly in all directions, and not bent over one way. Countersunk rivets require finishing with the flat end of a hammer.

Soldering

Soft soldering is a quick and useful method for making joints in light articles made from steel, copper, brass and aluminium, and for the conductor joints which occur in electrical work. Soldering itself does not make a strong joint. Mechanical strength is obtained by, say, marrying conductor strands or bolting busbars, and then soldering. Where joints are required to be mechanically strong, they should be riveted, brazed or welded. The subject of soldering is dealt with in Chapter 34.

Silver soldering is a hard soldering process which falls between soft soldering and brazing. These are two common silver solders:

Grade A: silver 61 per cent; copper 29 per cent; zinc 10 per cent. The melting range is 690–735 °C.

Grade B: silver 43 per cent; copper 37 per cent; zinc 20 per cent. The melting range is 700–775 °C.

A good soldered joint is recognised by a small amount of solder and perfect adhesion — rather than large unsightly blobs of solder. The following hints may be found useful:

1. Always use a bit with plenty of heat capacity.
2. A better joint can be made if the joint is warm rather than cold.
3. Iron tinning is made easier by having some blobs of solder in a tin lid with a little spirits, and touching both the spirits and the solder at the same time.

Brazing

The joining metal here is brass, which is a harder, stronger and more rigid metal than solder. As in soldering, an alloy of the brazing metal and the metal of the joint is formed at the surface of the joint metal. The brass used for making the joint in brazing is usually called 'spelter'. Its composition depends on the metal being brazed, because it is essential that the spelter has a lower melting point than that of the material to be joined. Spelter is obtainable in sticks or in a granular state, when it may be mixed with the flux (borax) before being applied to the joint. The heating for the joining process is obtained from a blow-pipe or blow-torch. The rules of cleanliness which apply to soldering alloy also apply to brazing; the work should be cleaned and polished at all points where the brazing is to occur. The work should be allowed to cool off normally, as quenching may lead to distortion of the joint or cracking of the spelter.

Chiselling

The cold chisel is an important cutting tool. An engineer's chisel does not have a wooden handle as do other types (e.g. woodworking). In the cutting process using a chisel, the tool should always be held about half way between the head and the edge, and at the correct inclination. If the angle is too great, the edge will cut too deeply. On the other hand, a shallow angle will not allow the chisel to cut efficiently. Generally, the smaller the angle the more effective are the hammer blows, and the more efficient the cutting operation. The edge should be kept well up against the shoulder formed by the cut and chip. Particles of metal should be kept away. When the chisel approaches the edge of the metal to be cut, particularly cast-iron, it should be reversed or the cut taken at right-angles to the previous one. Otherwise the edge of the metal is likely to be broken away.

Cutting with the hack-saw

The hack-saw is the chief tool used by the fitter for cutting-off, and for making thin cuts before chipping and filing operations. The choice of blade for any class of work is governed mainly according to the pitch of the teeth on the blade. These should be as large as possible to give the greatest clearance for the metal chips and to avoid clogging. Two at least, or three, teeth should always be in contact with the surface being sawn. Otherwise there will be a danger of the teeth being stripped from the blade. Also, if a corner is sawn too sharply, the teeth may get ripped off. Blades must be held tightly in the frame. Slow, firm and steady strokes (about 50 per minute) are best. On the return stroke the blade should be lifted slightly. Breakage of blades is caused by (*a*) rapid and erratic strokes; (*b*) too much pressure; (*c*) the blade being held loosely in the frame; (*d*) binding of the blade from uneven cutting; (*e*) the work not being held firmly in the vice. Solid metals should be cut with a good firm pressure; thin sheets and tubes need a light pressure. If the pressure at the start of a cut is insufficient, the teeth may glaze the work, and so cause their edges to be rubbed away.

Filing

To produce a flat surface by cross-filing is a difficult task: practice makes perfect in this activity as in others, and advice from skilled filers should not be ignored. The work should be held firmly in the vice, in a truly horizontal position. Grasp the file handle with the right hand, with the end of the file-handle pressing against the palm of the hand in line with the wrist-joint. The left hand is used to apply pressure at the end of the file. Strokes should be made by a slight movement of the right arm from the shoulder, and by a sway of the body towards the work. To get the movement right it is necessary to take up a stance to the left side of the vice, with the feet planted slightly apart. As the file moves over the work, it should be in an oblique direction, rather than parallel to its length. The file must remain horizontal throughout the stroke, which should be long, slow, and steady. Pressure is applied only on the forward motion. On the return stroke, although the pressure is relieved, the file remains in contact with the work.

Success in filing flat is dependent upon keeping the file really horizontal throughout its stroke. This position is controlled by the distribution of pressure between both hands. The fault with beginners is that too much pressure is applied on one hand, resulting in a round rather than a flat surface. Any tendency to rock should be corrected immediately. To test the surface of the work during filing, use a straight edge during the process and view the line of contact for 'daylight'. If a considerable amount of metal has to be removed, the bulk of the metal should be taken away by using a rough file. The surface should be progressively brought to a finish by using second-cut and smooth-files.

Draw-filing is the process used to remove file marks and to give a good finish to the work. For this purpose use a good fine-cut file with a flat face (e.g. a mill file).

Scraping

The purpose of the scraper is to correct slight irregularities from flatness. If these are great, a file should be used, for the scraping process is rather laborious and not intended for removing much metal. First, a standard of flatness must be obtained with which to compare the surface being scraped. A surface plate is used for this comparison. After thoroughly cleaning the surface plate, it should be smeared with a thin layer of prussian blue, or red lead in oil. The other surface (the work) should then be placed on it and rubbed about slightly. The high parts of the work surfaces will be smeared with part of the marking substance. These are the parts that must now be reduced by scraping. This is done by holding the scraper firmly to make short strokes 25 mm long. The scraper is left in contact with the surface for the return strokes, though no pressure is applied. Having gone over the surface with strokes in one direction, the next set of strokes should be made in the other direction, at right-angles to the first. This changing round procedure is repeated until sufficient metal is removed. From time to time trial rubbings should be made with the surface plate, and working on those parts which show 'high'. Gradually the surface of the work is brought to the condition that its whole area is covered by tiny areas of contact; it is usually taken as accurate if these spots are 3–5 mm apart. The scraper must always be kept sharp.

Drilling

If the work to be drilled is large and heavy there will generally be no danger of it moving or rotating dangerously, with the movement of the drill. Otherwise, with light work, always clamp or hold it securely by some method. Secure clamping ensures no danger to the drill operator, more accurate drilling and less breakage of drills. The chief danger in drilling is when the drill point breaks through at the underside of the part being drilled. When feeding the drill by hand, pressure should be eased off when the drill point is felt to be breaking through. Small drills should always be fed by hand. Special care is necessary when drilling thin plate because the drill point often breaks through before the drill begins to cut on its full diameter. The following hints are useful:

1. For soft metals use a drill with a quick twist to its flutes. The opposite applies for hard metals. For chilled iron, a flat drill gives the best results.
2. Cut with soluble oil for steel and malleable iron, and kerosene or turpentine for very hard steel. Cast-brass or iron should be drilled dry.
3. The blueing of a high-speed steel drill is not detrimental to it; but it is to a carbon-steel drill.

4. When drilling deep holes, release the feed occasionally and withdraw the drill. Remove metal chips from the bottom of the hole with an old round file that has been magnetised.

Marking out

Unless a particular job is simple, it is usual to indicate with marks where cuts, holes, slots and so on have to be carried out. The process of marking out means to use certain tools to prepare the job for some subsequent process to produce a finished piece of work.

The *marking-out table* has a metal (steel) top which is planed dead flat and is supported firmly. The *angle plate* is used for supporting a surface at right-angles to the surface of the table. It has plenty of holes and slots for taking the bolts necessary to secure pieces to it. The *adjustable angle plate* is used for angular work. *Vee blocks* are used to support round bars while they are being marked off. They are usually made in pairs. *Parallel strips* are used for supporting work on the marking-out table. Each pair of strips is of exactly similar dimensions, with parallel faces and all faces square. The *spirit level* is used for setting off surfaces parallel to the surface of the marking-out table. The level is also used for levelling machine beds and tables. A pair of *toolmaker's clamps* is an essential item in the fitter's kit; two or three pairs are the minimum number. They are handy items for clamping work, say, to an angle plate. *Dividers and trammels* are used for scribing circles, marking off lengths and so on. Trammels are used to extend beyond the range of dividers. Trammels with bent legs are sometimes used in the same way as inside calipers.

Feeler gauges consist of a series of blades or leaves of thicknesses varying from about 0.002 to 1 mm. They are useful for flatness tests. Other tools for marking off with accuracy include the *try square* for testing squareness, *protractors* and the *bevel* for testing the angle between two surfaces.

For any scribed line to be visible it is first necessary to prepare the surface which is to receive the line. The rough faces of castings which have to be marked should be brushed over with a little whitewash which, when dry, will show up the scribed line. Machined surfaces may be prepared by brushing them over with copper sulphate solution which leaves a thin film of copper on the surface; this shows up a scribed line very clearly. It should be noted that marking out does not mean that one can dispense with the use of measuring instruments for the control of machining and cutting processes. Always the line should be used as a guide, the final finishing being done to micrometer, vernier, or some other exact measuring method. Chapter 32 deals with workshop measurements.

Drilling and tapping a hole

For tapping it is first necessary to drill a hole the diameter of which must be approximately equal to that of the bottom of the thread. When the hole to be tapped is 'blind' (i.e. open at one end only) it should be drilled one or two threads deeper than is required for the finished depth of the tapped hole.

Having drilled the tapping hole, the taper tap is secured in the tap wrench and started in the hole. Before beginning to screw it round for cutting the thread, its position must be adjusted until it stands square with the top surface of the work. This can be done by setting the tap vertical with the aid of a small square. Enter the tap and keep it straight for the first few turns. A little oil will help the action of the tap and improve the finish of the threads. When the taper tap is felt to have started cutting, and its squareness checked, the cutting of the thread proper may proceed. The tap should not be turned continuously, but it should be reversed about every half-turn to clear the threads of metal particles. If any stiffness is encountered (this can be felt on the tap wrench), no force should be used, but the tap carefully eased back and fore to clear the obstruction. When a blind hole is being tapped, the tap should be withdrawn from time to time and the metal cleared from the bottom of the hole. If the hole is a straight-through one, the reduction in resistance will indicate that the taper tap is cutting the full thread. It can then be withdrawn and replaced with a second tap. When a blind hole is being tapped, increase in resistance will be felt as the point of the tap reaches the bottom of the hole. No force must be used at this point as the tap may be broken or the threads stripped. Remove the taper tap and take a second tap down as far as it will go. Remove this tap and replace it with a plug tap. Great care should be exercised when using small, slender taps, particularly in blind holes, as the tap may break inside the hole. If this happens, the piece may be removed with a punch. Otherwise, it will have to be softened by heating, drilled out and the hole retapped.

Threading

The tool used for cutting external threads on bars and tubes is called a die. Basically it consists of a nut with portions of its thread circumference cut away and shaped to provide cutting edges to the remaining portions of the thread. The die is used with a pair of operating handles called stocks. The action of dieing is similar to that of tapping, except that it is more difficult to keep a die square. It is therefore necessary to exercise great care in threading. When the cut has begun, the die should be worked back and fore similar to the method used for tapping. Certain designs of stocks and dies incorporate guides which ensure that the die is kept square-on. The conduit stocks and dies for the electric thread show an example of this provision.

Grinding

This process involves the use of abrasive wheels to cut, surface and shape materials. The most common type of wheel is the straight grinder, and is generally used in the workshop for sharpening tools and the rought shaping of a surface before a final finishing process. The main points to bear in mind include the wearing of goggles to protect the eyes, and the holding of the work piece securely.

Fitting pulleys to motors

First ensure that both the motor shaft and the pulley bore are quite clean. The key should be fitted into the shaft, taking care to see that there is clearance between the top surface of the key and the bottom surface of the pulley keyway when the pulley is in its final position. Thus, the key should fit only on the sides of the keyway. The shaft and bore should then be lightly smeared with lubricating oil and the pulley worked onto the motor shaft. Ensure that the shaft key is in line with the pulley keyway. The pulley should be driven home by light taps with a hammer on a piece of wood held against the pulley hub. The blows of the hammer should be distributed round the whole surface of the hub to ensure uniform pressure. The pulley should be driven right home and any set-screw firmly tightened onto the key.

Extraction of pulley

This task is best done with the use of the tool called 'pulley drawers'. It consists of a pair of jaws coupled together at one end and containing a long threaded bolt which, when turned, exerts pressure on the motor shaft and so draws off the pulley. A small piece of copper or other soft metal should be placed between the point of the screw and the shaft end to avoid unnecessary damage.

Overhauling

The following is a summary of the procedure to be followed when overhauling a motor in the electrical workshop:

1. Dismantle the motor carefully; do not open cartridge bearing housings.
2. Clean away all dust, dirt, oil and grit by using a blower, compressed-air hose, bellows or brushes, and with petrol if necessary, or carbon tetrachloride (CCl_4). The complete removal of any foreign matter is important.
3. Check all parts for damage or wear, and repair or replace as found necessary.
4. Measure the insulation resistance and dry out windings if necessary until the correct value is obtained. Replace or repair any damaged windings.
5. Reassemble all parts. Ensure that machine leads are on the correct terminals and that all parts are well tightened and locked, where this provision is made.
6. Check insulation resistance.
7. Check air gaps.
8. Commission and test.

34 Soldering

The process of soldering in electrical work is generally used to make joints in conductors and to terminate a conductor for a mechanical connection to an appropriate terminal in electrical equipment. With the emphasis nowadays on speed, while still producing a good job of work, many aids to jointing have been produced. For instance the married joint which was once used to make a straight-through connection between the ends of two cables is now being replaced by the weak-back ferrule. And the tee joint is being replaced with the use of the claw-type clamp.

The bulk of information relating to conductor joints and terminations is contained in Chapter 7 of this book. This chapter deals with the subject of soldering as a workshop practice, and actual jointing procedure is only touched on where relevant.

The nature of soldered joints

The essential feature of a soldered joint is that each of the joined surfaces is wetted by a film of solder, and that the two films of solder are continuous with the solder filling the space between them. When solid, the joint becomes a hard mass of the material to be joined and the solder.

Basically, solders are metallic substances which have lower melting points than the metals they are to join together. They act (*a*) by flowing between the metal surfaces to be joined; (*b*) by filling completely the space between the surfaces; (*c*) by adhering to the surfaces; and (*d*) by solidifying. Once it adheres to a metal surface, solder cannot be prised off, because it becomes attached chemically to the surface and forms an intermediate compound. Thus, the soldered joint is, if made correctly, a most effective method of producing an electrically and mechanically sound form of connection between two current-carrying conductors.

The basic steps in soldering

Making a soldered joint can be divided into five basic steps:

1. Shaping the metal parts so that they fit closely together.
2. Cleaning the surfaces to be joined with a special substance.
3. Applying the soldering flux.
4. Applying molten solder.
5. Removing surplus solder and cooling the joint.

The surfaces to be joined should fit so that the space between them is narrow enough to become completely filled with molten solder (drawn in by capillary action as the solder wets the metal). If there is not enough clearance between the surfaces, the solder may not be able to penetrate and the result may be a joint with holes.

The reason for cleaning the surface is to expose the bare metal and make it free from grease or oxide which would otherwise prevent the solder from adhering to the metal. Rough cleaning consists in filing, scraping or rubbing with an abrasive cloth or paper. However, to get the surfaces really clean it is necessary to apply some substance which will remove any remaining particles of foreign matter. For large jobs degreasing is done by a solvent application or by cleaning in a bath of weak solution of an alkali (sodium salts) in hot water. In smaller jobs, such as jointing the smaller sizes of conductor, the flux used is sufficient to clean the surfaces of the metals to be joined together.

When the metal surfaces are being heated prior to the soldering operation, oxides may form on them. To prevent this happening a flux is applied. The basic characteristics of any flux are:

(*a*) It should be a liquid cover over the metal

and exclude all air.

(*b*) It should function as (*a*) while the surfaces are being heated up to soldering temperature.

(*c*) It should dissolve any oxide that may have formed on the metal surfaces, or on the solder itself, and carry these away.

(*d*) It should be easily displaced from the surfaces once the molten solder is applied.

Fluxes for electrical work must not remain acidic or corrosive at the completion of the soldering operation. Thus, acid fluxes such as 'killed spirits' should not be used, because they tend to cause corrosion in the joint after soldering. The fluxes used in electrical work include pure amber resin (or rosin) in some form, 'activated' resin as used in cored solders, and solder paste. Resin is the gum exuded from artificially produced wounds in the bark of pine trees. At ordinary temperatures it is solid and does not cause corrosion, but it reacts mildly at close to soldering temperatures. Resin is easily crushed to a powder form and melts readily at 125 °C. It can be applied by sprinkling the powder on the joint, but the more effective method is to brush it on to the joint as a solution. The usual solvent for resin is methylated spirit or industrial spirit. Activated resin fluxes contain a chemical which reacts with the oxides on the heated surface of a metal to clear them away, but they do not remain corrosive after the soldering operating.

Solder pastes are used where liquid fluxes would drain away too quickly from the surfaces to be joined.

The next step in the soldering process is the tinning which, in the case of tinned-copper conductors, has been done already. This process is the spreading of a thin layer of solder over the surfaces to be joined so that the surfaces are 'wet'. If the solder does not spread quickly and easily, then this indicates some foreign matter in the area (either oxide or grease). If this is the case, the surfaces should be cleaned with the application of a fluxed cloth-pad, more flux and then retinning. Without the wetting of the metal surfaces there can be no soldering action. The molten solder should leave a continuous permanent film on the surfaces instead of rolling over them.

Once the joint surfaces are thoroughly wetted and the space between them is completely full of solder, it remains only to cool the joint after wiping off any surplus solder using the fluxed cloth-pad. Cooling should in most instances be natural. Rapid cooling may cause a 'dry' joint which has a high resistance — the result of the soldering cracking.

During the soldering process it is essential to ensure that the job is clamped securely and that the clamping device holds the parts to be joined firmly and accurately.

The soldering bit

The function of the soldering bit is to store and carry heat from the heat source to the work, to store molten solder, to deliver the molten solder and to withdraw surplus molten solder. As a carrier of heat the bit should have a large heat-storage capacity. The size of the bit for any soldering job is determined by the rate at which it has to supply heat to the work. The material for the bit is most often copper, which wets readily and to which molten solder clings without difficulty.

Various methods are used for heating the bit. In days gone by the source of heat was a stove burning charcoal or coke; coal fires were not favoured because of soot. Nowadays butane gas and electric heating are used.

Care of soldering bits

Soldering bits rapidly become coated with scale (oxide) and the faces that are wetted by flux or solder become pitted. Pitting develops during the life of the bit, sometimes quite early in its life. The result of a pitted bit is that the work becomes impaired and it becomes necessary to file the faces of the bit until a sufficiently large surface area is obtained and tinning becomes easy. Pitting is the result of (*a*) oxidation of the copper due to heat, (*b*) attack by the flux on the copper face and (*c*) the transfer of copper particles from the bit into the solder. Care involves the maintenance of the soldering temperatures and no higher. Excessive temperatures result in whole-scale oxidation and rapid wear. At all times the shape of the bit should be preserved. Each bit should be of a shape appropriate to the job.

Wiped joints

A wiped joint is a full-bore straight connection between two pipe-like sections. In electrical work one section is the lead sheath of a cable; the other section is the entry to a cable sealing box, or a straight-through or tee-joint box. Making a wiped joint is a skilled task and is usually performed by experienced cable jointers.

There are two methods:

(*a*) The first is the blowlamp method, though a blow-torch is more often used. The area to be jointed is heated by playing the flame of the torch over it. A solder stick is held in the flame and, when the surface is hot enough, is applied round the joint in order to tin it. The solder is rubbed over the joint area which has been previously prepared with flux (tallow is often used). Additional solder is melted onto the area and is caught on a wiping cloth made from several layers of buckskin cloth. The molten solder is worked round and round to ensure an even consistency in the plastic mass. When enough solder has been applied, the solder stick is laid aside and the mass of plastic solder melted off onto the cloth. The cloth, of course, must always be kept smeared with tallow and heated on its working face.

The mass of the solder is then lifted back round the joint and wiped into shape without the flame. The molten pasty solder must be kept moving and the shaping completed before any indications of setting are seen. This is skilled work which requires experience to do a good job. This method is suitable for horizontal working.

(*b*) The second method of making a wiped joint is suitable for vertical joints and is known as the pot-wipe method. One advantage of this method is the reduced risk of overheating the lead sheath. The joint is heated by dribbling molten solder over it from a ladle. The solder is melted in a pot. At first the solder is chilled by the coldness of the joint area and solidifies around it without adhering. As the pouring is continued the mass gathers heat and becomes pasty. The pipe surface is raised in temperature and becomes tinned as the solder melts off the pipe to be caught in the wiping cloth. The wiped joint is formed as already described. A circular catch-plate of sheet metal with a central hole is fixed round the lead sheath. To prevent solder from adhering to the sheath beyond the required limit on each side of a wiped joint, plumbers' black is used and is applied beforehand.

About solders

Two basic types of solder are used in electrical work: fine solder (tinman's solder) which is 60 parts tin and 40 parts lead, and plumber's metal which is 30 parts tin and 70 parts lead. Fine solder melts more easily and so is more commonly used for electrical joints. Fine solder is used for conductors; plumber's metal is used for wiped joints because it remains in a plastic state longer than the finer metal.

Solder is made in a variety of forms: bars and sticks, and in wire and cored forms. Wire solder is convenient to use and comes in various diameters. For fine electrical and electronic work finer gauges of wire are available. These wires melt instantly at a touch of the copper bit and their uniform gauge makes it easy for just the right amount of solder to be applied to a joint. Cored solders are hollow or grooved wires or tubes filled with flux. The flux is introduced warm as a paste during the drawing of the tube, and solidifies when cold. With the flux in the cored solder some considerable time is saved when a number of small joints have to be done quickly. Solders are also available in paste, cream and powder forms.

Soldering aluminium

For many years it has been accepted that aluminium cannot be soldered easily or satisfactorily by methods commonly used for soldering copper. The oxide film which forms with heat on the surface of aluminium is very resistant to chemical attack, and although the film can be removed by mechanical means, the cleaned surface thus exposed will oxidise very rapidly and the abrasive method of cleaning is, therefore, a very uncertain one. This problem has been solved by the development of new fluxes which will cope with this oxide film and it is now possible to solder aluminium by the conventional methods used in soldering other non-ferrous metals. The fluxes are available in paste and liquid forms and incorporated in cored solder-wire. The solders are in stick form.

Soldering aluminium today is now a straightforward process. The following points must be observed:

(*a*) All surfaces must be scrupulously clean.
(*b*) When a joint is being made between stranded conductors the strands must be 'stepped' to increase the surface area.
(*c*) The surface must be heated before the flux is applied. The flux will take only when the temperature is high enough.
(*d*) Apply aluminium solder until the surface is bright.
(*e*) Joints in aluminium should always be protected from contact with the atmosphere: by painting, taping or compounding.

Index